T0219751

Learning Engineering Practice

Learning Engineering Practice

James P Trevelyan

CRC Press
Taylor & Francis Group
Boca Raton London New York

CRC Press is an imprint of the
Taylor & Francis Group, an **informa** business

Cover image: City of the Arts and Science, Valencia, Spain.
© Zebbache Djoubair unsplash.com

CRC Press/Balkema is an imprint of the Taylor & Francis Group, an informa business

Library of Congress Cataloging-in-Publication Data
Names: Trevelyan, James P., author.
Title: Learning engineering practice / by James P Trevelyan.
Description: Boca Raton : CRC Press, [2021] | Includes index. | Summary: "This book explains engineering practice, what engineers actually do in their work. The first part explains how to find paid engineering work and prepare for an engineering career. The second part explains the fundamentals of engineering practice, including how to gain access to technical knowledge, how to gain the willing collaboration of other people to make things happen, and how to work safely in hazardous environments. Other chapters explain engineering aspects of project management missed in most courses, how to create commercial value from engineering work and estimate costs, and how to navigate cultural complexities successfully. Later chapters provide guidance on sustainability, time management and avoiding the most common frustrations encountered by engineers at work. This book has been written for engineering students, graduates and novice engineers. Supervisors, mentors and human resources professionals will also find the book helpful to guide early-career engineers and assess their progress. Engineering schools will find the book helpful to help students prepare for professional internships and also for creatingauthentic practice and assessment exercises"—Provided by publisher.
Identifiers: LCCN 2020031410 (print) | LCCN 2020031411 (ebook) | ISBN 9780367651794 (hardback) | ISBN 9780367651817 (paperback) | ISBN 9781003128212 (ebook)
Subjects: LCSH: Engineering—Practice. | Engineering—Vocational guidance.
Classification: LCC TA157.T765 2021 (print) | LCC TA157 (ebook) | DDC 620.0023—dc23
LC record available at https://lccn.loc.gov/2020031410
LC ebook record available at https://lccn.loc.gov/2020031411

Published by: CRC Press/Balkema
Schipholweg 107C, 2316 XC Leiden, The Netherlands
e-mail: Pub.NL@taylorandfrancis.com
www.crcpress.com – www.taylorandfrancis.com

ISBN: 978-0-367-65179-4 (hbk)
ISBN: 978-0-367-65181-7 (pbk)
ISBN: 978-1-003-12821-2 (ebk)

DOI: 10.1201/b22622
https://doi.org/10.1201/b22622

Typeset in Times New Roman
by codeMantra

Contents

Preface

This book is being published as nations around the world struggle to recover from the consequences of the SARS-CoV-2 virus pandemic. Governments have imposed movement restrictions to slow the spread of infections so that hospitals have enough capacity to provide care for the significant minority of infected people who develop acute breathing difficulties. As a result, many industries such as commercial aviation, hotels, retail stores, manufacturing, and entertainment have been forced to close or drastically restrict their operations. Governments with sufficient economic capacity have provided payments and subsidies for affected workers. In low-income countries, tens or hundreds of millions of low-paid and migrant workers lost their incomes and have been forced to leave cities for their ancestral villages in the hope of finding food and shelter. The world's economies, therefore, are in an unprecedented state of disarray, confusion, and uncertainty.

All this means that young engineers – students, recent graduates, or novices in the first stages of their career – face enormous uncertainties. Students may have difficulties finding internships to get engineering work experience needed to complete their degrees. Graduates may find it much harder to find paid engineering work in their fields, and many novice engineers will find themselves unemployed as engineering firms encounter tough trading conditions.

Even before the pandemic, engineering performances around the world had been disappointing governments and investors alike for two decades. Productivity is a macro-economic measure of value produced by firms relative to the human effort expended (labour productivity) and other inputs such as capital invested, research and development, energy, land and materials (multifactor productivity). Engineering has a crucial influence on productivity. Engineered tools and equipment amplify human effort and increasingly mental capacity as well. Capital is invested in engineering projects, much of the research and development is performed by engineers, and engineering enterprises provide materials and energy. Productivity growth around the world has slowed significantly since about 2005. For example, US labour productivity increased by nearly 1% annually through previous decades, but since 2007 has improved by only 0.5% in total!

The five-times productivity gap between wealthy countries and low-income countries has hardly changed in the last sixty years, despite the wide availability of technological advances. What does that gap mean in practice? In simple terms, on average, most material goods and engineered services of equivalent quality and durability are many times more expensive in low-income countries. For example, electrical energy at the point at which it performs useful work can be five times more expensive. In most low-income countries, electricity supplies are intermittent so standby generators are needed to ensure continuous supply, greatly increasing the cost. Cheap, low quality appliances require more electrical energy to provide the same useful work. Low quality maintenance further degrades the efficiency of energy conversion.

At the same time, almost every facet of our human civilization needs to be reconceived, redesigned, rebuilt, refurbished, or at least re-arranged, to ensure sustainability. In other words, productivity has to be improved enormously so that people can enjoy the benefits of our technological civilization while using far less material resources and energy, and reducing greenhouse emissions to near zero, all in the next three decades.

Therefore, despite the apparent uncertainty and the difficulties finding work in the next two or three years, especially for young engineers, the opportunities have never been greater than now.

It is important to understand that engineering schools mainly teach engineering science: written knowledge based on mathematics and physical sciences, increasingly life sciences as well. While this knowledge is critical for all engineers, it is only a foundation. Engineers are needed by firms to put that knowledge to practical use, to deliver working solutions that meet the needs of people who ultimately pay for the them. Investors provide the money to pay for these engineering solutions to be built and delivered, long before the people who use them start paying. Therefore, it is engineers who have to reassure company directors that the solutions will work as intended, to build the confidence needed for the necessary finance to be raised from investors. These developments, in turn, create jobs and opportunities for so many other people. Therefore, practical engineering capability is an essential starting point for economic recovery and the necessary transformation of our human civilization.

This book serves as a guide for young engineers to develop the practical capabilities that engineering schools cannot teach, mostly by learning to collaborate effectively with all the other people involved in an engineering enterprise.

The book emerged from research on engineering practice, studying the working days of engineers from many different disciplines and settings, in wealthy and low-income countries. It reflects contributions from colleagues, students, and many other researchers. My earlier book *The Making of an Expert Engineer* explained many of the research results in detail. I wrote it for two audiences: first, practising engineers and, second, faculty staff in engineering schools who teach engineering.

In recent years I have focused on the challenges that early-career engineers face as they transition to the realities of engineering workplaces. Many face frustrations and disappointments as they encounter seemingly mundane tasks that seldom, if ever, demand the advanced mathematics and scientific capabilities they developed through their university education.

I have continued to supervise young engineers, reread many of the hundreds of interview transcripts, and listen to comments from frustrated employers. From this I realised that engineering students develop deeply held expectations and values that arise from the way that contemporary formal education is structured, irrespective of the course content provided by engineering schools.

Students learn to value independent thinking, creating and writing solutions for well-defined intellectual technical problems presented in writing, and framed in terminology that constrains interpretation. They expect to find employment that requires them to create solutions for challenging technical problems.

These expectations and values contrast starkly with workplaces characterised by interdependent work that demands collaboration, an ability to listen carefully, and engage in complex social performances with people who have diverse interpretations of expectations and requirements. Economic and human constraints, more so than technical limitations, dominate practically all engineering workplaces. While students learn to value written communication, they enter a world where oral communication can be far more important and is inevitably refracted by beliefs and coloured by emotions.

Eventually, I realised that I should use this research to write a book that could help novices learn to practice engineering faster than they do now, more or less by experience alone.

I am confident that this book will help a new generation of engineers rise to the challenges we all face together.

James P Trevelyan, May 2020

Author biography

Emeritus Professor James P Trevelyan is an engineer, educator, researcher, and recently became a start-up entrepreneur.

He is CEO of Close Comfort, a tech start-up introducing new energy-saving, low-emissions air-conditioning technology to Australia, Indonesia, Pakistan, and other countries with a large potential global market.

His research on engineering practice helped define *Engineers Australia* professional competencies for chartered engineers. His book *"The Making of an Expert Engineer"* and advances in understanding how engineers contribute commercial value are influencing the future of engineering education in universities and workplaces. Another book, *"30 Second Engineering"*, is helping to build greater awareness of the key importance of engineering and will reach a global audience.

He is best known internationally for pioneering research that resulted in sheep-shearing robots from 1975 till 1993 and for the first industrial robot that could be remotely operated via the internet in 1994. He received the leading international award for robotics research, equivalent to the Fields medal in mathematics.

In 2018, he was awarded West Australian of the Year in the professions category in recognition of his achievements.

Web pages:
https://www.closecomfort.com/
https://JamesPTrevelyan.com/
https://research-repository.uwa.edu.au/en/persons/james-trevelyan
https://www.linkedin.com/in/jtrevelyan/

Acknowledgements

My dear wife Samina has supported me throughout decades of research and writing.

Her father, Malik Muhammad Iqbal Khan, first opened the doors to understanding engineering in Pakistan.

Her mother, Begum Sarfraz Iqbal, helped me understand how to navigate countless aspects of South Asian culture.

My family and The University of Western Australia supported the research financially, with contributions from scholarships and several companies. Colleagues such as Sabbia Tilli, Prof. Sally Male, and Melinda Hodkiewicz all supported this research in different ways. My students, especially my PhD students, contributed research evidence and raised many pertinent questions.

Hundreds of engineers freely gave their time to be interviewed, observed, and questioned. Some read ideas emerging from the research and provided comments and feedback.

Practising engineers Mario Dona, Richard Bercich, Anand Jyothi, Warwick Bagnall, Robert Harrington, Catherine Morar, Roger Penlington, and Zhigang Ji all read early draft chapters and provided insightful feedback and suggested improvements.

Michele Thomas provided valuable suggestions to help engineers on the autism spectrum.

My editor, Marli King, greatly improved the flow and readability of the text.

Part I

Preparation for an engineering career

Engineering is not what you might expect

Engineering students eagerly look forward to highly technical design and analysis work requiring elaborate software tools and solving technically challenging problems.

However, many graduates have a frustrating time trying to find paid employment and send out countless job applications, often without a response or acknowledgement. Even finding unpaid internships can be challenging.

When they eventually start working, graduates often find themselves assigned to seemingly simple and mundane tasks. Little can be achieved without the collaboration and coordination of other people who may seem preoccupied and disinterested. The advanced analysis capabilities acquired in their education may seem almost irrelevant.

Students today are silently conditioned to think that independent, intellectual, written work is the key to success because exams test that ability.

Engineering, instead, demands collaboration with other people from a wide range of backgrounds, from finance and trades to marketing and sales. Success in engineering depends on working with technical, business, and social factors that are all intertwined with each other.

Recent research has helped to clarify the principles upon which a successful engineering practice depends. Collaboration requires much more than conventional teamwork, as technological expertise demands specialised techniques, many of which are explained in this book. It takes time and effort to learn and practice these new skills.

This book provides a curriculum based on this research to guide workplace learning for the first 3 years of an engineering career, enabling novices to practice independently. The book also explains how to acquire critical knowledge that today is mastered by just a few engineers.

Is there a particular personality or set of abilities that is perfectly aligned with engineering? No! The ideas in this book demonstrate that there is a place in engineering for everyone with appropriate persistence and qualifications. Everyone needs help from others to succeed; this book will help you find the support you need to make the most of your abilities as an engineer. The world needs your enthusiasm, ideas, and contributions!

This book is based on 20 years of systematic research. The insights are derived from interviews with hundreds of practicing engineers, along with field observations from many engineering firms and projects in several countries. This research was a joint effort with many students and valued colleagues.

There is also specific guidance for working in low-income countries included in the text. One of the most exciting discoveries from recent research is that the social, cultural, and economic environment strongly influences engineers' performances. This insight can help engineers in low-income countries navigate the socio-cultural complexities that frame their daily practice. This guidance has the potential to greatly improve enterprise productivity, creating enormous social and economic improvements in low-income countries.

Chapter 1 describes engineering and the challenges ahead. Chapter 2 explains engineering practice and how the book should be used with the Professional Engineering Capability Framework document to guide and record all learning progress.

Chapter 3 explains the essential first step: finding paid engineering work.

Chapters 4–8 cover critical perception skills that are commonly neglected in formal education, with pointers on how to improve them while finding work. Chapters 9–20 explain the fundamentals of engineering practice.

Anyone who wants to boost their career prospects should go on to read my earlier book *The Making of an Expert Engineer*. It can help experienced engineers take their practice to more advanced levels.

Additional material for this book can be found at the book page on the publisher's website: https://www.routledge.com/9780367651817.

The author's webpage also provides additional material, enabling readers to access the Professional Engineering Capability Framework and other online supplements for both novice engineers and their supervisors at https://www.jamesptrevelyan.com/.

Chapter 1

Engineering: doing more with less

What is engineering? It can be a mysterious occupation. Many people imagine engineers design and perform complicated mathematical calculations. Some engineers do that, but very few spend much time on it. Others think that engineers build bridges or make cars. However, few engineers would know how to make or even fix a car. If you see an engineer working with tools on a bridge, something is probably very wrong. Get out of the way—fast! (Figure 1.1)

Engineering is much more than what engineers do, although the path to understanding it is to understand engineering practice, which is what engineers actually do. Recent research has greatly expanded our knowledge of engineering practice, revealing that practically all engineers use the same ideas and methods introduced in this book.

Engineering is a knowledge-based profession. So, what does that mean?

In essence, engineers are people with specialised technical knowledge and foresight who conceive, plan, and organise the delivery, operation, and sustainment of man-made objects, processes, and systems. These engineered solutions enable people to

Figure 1.1 An engineer in the popular imagination: a man with a hard hat, working with a drawing. Images like this portray misleading stereotypes because real engineering has largely been a mystery... until now. (Photo: raxpixl at unspash.com)

be more productive: to do more with less effort, time, materials, energy, uncertainty, health risk, and environmental disturbances.

Most engineers organise their work into projects.

Project definition starts with engineers conceiving safe solutions for human needs, often helping clients better understand their needs and solutions in terms of engineering possibilities.

Working mostly with computers and simulations, engineers predict how well these solutions will work, as well as the cost to build, operate, sustain and, eventually, remove them. Engineers often predict the commercial benefits of these systems for customers and end-users. However, there are always uncertainties, so engineers also inform clients and investors about risks and consequences. Sufficient trust and confidence must be built before clients or investors are willing to provide financial support, long before any benefits from a project begin to arise. All of this work leads to a decision by investors to proceed with project execution, the second phase (Figure 1.2).

In the execution phase, engineers plan, organise, and coordinate the collaborative efforts of skilled people, guided by shared knowledge, to construct the chosen solution. Much of the effort is needed to ensure that the original intentions are implemented faithfully enough to achieve the expected technical and commercial performance. This is time-consuming work that usually involves many people. Solitary technical work, such as performance prediction, design, and solving technical problems, takes up much less time.

Today, engineers often collaborate in large teams with people in different parts of the world and different time zones. They plan, organise, and teach people to purchase and deliver components, tools, and materials, and then transform, fabricate, and assemble them to deliver the intended solution. They work with an agreed schedule and budget, handling countless foreseeable but unpredictable events that affect

Figure 1.2 Engineers in a consulting firm discussing vibration measurements collected from a natural gas processing facility. Their client, a natural gas pipeline operator, requires assurances that there is an acceptably low risk of failure. Each of the engineers in this meeting has different expertise and experience, and they are utilising their collective expertise to arrive at a shared conclusion.

progress, performance, safety, or the environment. Later, they organise sustainment: operations, upgrades, maintenance, and repairs.

In the final phase, engineers plan and organise removal, disposal, and environmental restoration. Materials are often reused or recycled, while specific components are often refurbished and sold.

The ultimate objective is usually to satisfy client, investor, and end-user expectations well enough that investors will return and commission more projects.

In this way, engineering success stories almost always reflect the contributions of dozens (or even hundreds) of engineers and thousands of other people worldwide, building on decades of experience. They also reflect specialised ways for all these people to collaborate and share technical understanding, using methods that have evolved over centuries of practice, often encoded in organisational procedures.

Certainty is impossible with unpredictable activities by so many people. Natural variations in materials and the environment add even more uncertainty. Yet engineers have evolved systematic methods that provide amazing predictability. Few people watching the hair-raising exploits of aviation pioneers in the early 20th century could have imagined the amazing safety and reliability of modern air travel.

Engineering, therefore, is all about specialised technical knowledge built on science and mathematics, on the one hand, and people who need to use and apply that knowledge, on the other hand. Engineers have to work with the limitations of both people and specialised knowledge. Performance is constrained as much by human limitations as by the laws of physics. The value of a product depends as much on the buyer's subjective perceptions, knowledge, and ability to use it effectively as it does on the original design and manufacturing quality. Most engineers graduate with a good understanding of engineering science, but little if any understanding about people and human limitations. That's why this book focuses more on the human aspects of engineering practice, particularly on effective collaboration methods.

Invention and innovation are highly valued in engineering. Even though relatively few engineers get the chance to work on leading-edge technologies or research and development, these are often seen as 'dream jobs.' Engineers love the intellectual challenges that innovation presents. In practice, innovation is always tempered by accumulated knowledge and standard methods shaped by past experience. Hundreds of company, national, and international standards embody this breadth of experience. Gradually, engineers come to realise that even the most apparently mundane engineering jobs present fascinating challenges, and the most complex challenges are invariably about people.

Another guiding principle is the ethical responsibility for efficacy and safety of engineered solutions: not to cause loss, harm, or suffering and to avoid wasting resources. Engineers tend to honour ethical obligations for practical reasons because effective collaboration is based on trust from clients, contractors, and employees. They work in small communities, so news of a breach spreads fast. Ethical principles are articulated in professional codes of ethics, as well as government and social regulation dating back to the Code of Hammurabi thousands of years ago. These principles help to shape conscientious efforts by engineers who produce many of humanity's most durable achievements.

Trust from government, regulators, and local communities always helps. Earned over time, companies build a 'social licence': respect and trust accumulated through consistent and ethical performances that meet regulatory and societal expectations.

Engineering tends to be a culturally diverse, male-dominated profession. Women are gradually becoming more numerous, especially in fields like biomedical, environmental, food processing, and chemical engineering. Several Muslim countries have some of the world's highest female participation rates in engineering. Many firms now recognise that diversity is critical for finding solutions that address all the required issues quickly, and they are now actively seeking to recruit and retain female engineers and engineers from diverse cultural backgrounds. Since so much engineering work operates across national borders, having staff familiar with foreign languages and cultures makes the work far easier.

It is not unusual to see predictions that computers with artificial intelligence (AI) will eventually perform many of today's engineering jobs. So far, however, most of these forecasts have been wrong. Predictions since the 1970s that robots and AI would eliminate factory work, for example, have proved to be premature. That being said, AI has improved information technology (IT) system performance, enabling engineers to find appropriate information much faster. AI can also help robots perform more consistently, and computer systems, often with AI components, do help to enormously extend human capabilities.

Most engineers love their work and often enjoy the thrill that comes from spending large amounts of other people's money, overcoming challenges with ingenious solutions and transforming technical ideas into reality to generate great benefits.

Today, and for the foreseeable future, engineers provide essential leadership for many of the most significant advances in human civilisation. Some recent achievements include IT and communications, extending the internet across the planet and into space. Through their work on water supplies, construction, sanitation, and transport, engineers have enabled human populations to grow and occupy almost all the dry land on the planet. Tentative steps are being taken to extend human presence to the ocean's depths and even into outer space.

Transforming the planet

Now, the greatest immediate challenge facing all engineers is to reduce resource consumption, environmental disturbances, and greenhouse emissions to ensure that human civilisation is sustainable in the long term within the limitations of the planet's biosphere.

We all know that we're consuming too much of the world's resources. While a minority lives in comfort, it is only reasonable for the poor majority to aspire to the same living standards. Yet, if everyone on the planet consumed resources at the same rate as the comfortable minority, the remaining resources would quickly be exhausted. We need to do much more with much less, and soon; in other words, we need to boost productivity!

In most emerging markets, engineering products and services are much more expensive for equivalent quality in all respects. This seems counter-intuitive: most people think that engineering should be much cheaper in a country with low hourly labour rates.

Take safe drinking water, for example. In Australia, the driest continent, copious amounts of potable water—clear, clean, and safe to drink—flow from kitchen taps 24 hours a day at a total cost of about USD 3 per tonne. In emerging economies like India, Pakistan, and Indonesia, potable water often costs between USD 20 and 100 per tonne. Getting the bare minimum needed for survival may demand one-quarter of the economic resources of low-income families, representing more than 10% of the GDP in a country like Pakistan.

Engineers can change that, bringing huge improvements for billions of people who currently live in poverty (Figure 1.3).

This is what frames my life today as an engineer: leading Close Comfort, an Australian start-up creating new air-conditioning technology. Today, only 2%–3% of people in countries like India can routinely enjoy air conditioning. Everyone living in hot and humid regions of the planet needs air conditioning to be reasonably comfortable while sleeping and working. Yet the energy needed for billions of people to enjoy 20th-century air conditioning is unimaginable, and we would literally cook the planet. We are developing new technology to enable almost everyone to enjoy air conditioning, using renewable energy with minimal environmental effects, and without significant modifications to existing buildings. We can transform air conditioning in 20–30 years, soon enough to avoid unacceptable environmental damage as we phase out older technology. It's a sustainable solution that also saves money for users.

We did not set out to improve the efficiency of the conventional room air conditioner.

Close Comfort is an entirely different concept that began with my first-hand experience of the need for it.

It was around 2 AM in my first-floor bedroom in Islamabad, Pakistan, early in June 2004. The national electricity grid was unable to keep up with the power demanded by too many room air conditioners, so the government introduced load shedding. Electricity was cut off for an hour or so in each neighbourhood in turn. The old window air conditioner thudded to a halt, and the ceiling fan slowed to a halt. I lay in complete stillness and total darkness. Within minutes, the room temperature was about 40°C, sweat was running down my brow, and I could hear the mosquitos, ready to attack.

Surely, I thought, a tiny battery-powered air conditioner could keep a flow of cool fresh air on my face and neck inside a mosquito net; that would provide enough comfort for me to sleep.

Years later, Close Comfort is a commercial product running on 300 W, a low enough electric power demand to run on domestic UPS inverters—the standby battery power supplies found in homes across South Asia, Africa, and many other regions with intermittent electricity supplies. In essence, it is a tiny fridge with a fan inside to blow cool air at people. The idea is to cool people, not rooms: a personal air conditioner.

Figure 1.3 Close Comfort personal air conditioner.

Close Comfort also illustrates another key principle in engineering—KISS: Keep It Simple, Stupid. It is a simple solution, but achieving elegant simplicity in engineering is not easy. Close Comfort required years of patient research and experimentation. Achieving simplicity can be extremely challenging, but also extremely satisfying when something so simple works so well and brings pleasure to so many people who use it.

As engineers, we can transform all aspects of human civilisation. No matter where you, the reader, are working right now, we can all play a part in boosting productivity and sustainability. While improving energy efficiency is part of the solution, we also need completely new approaches, such as Close Comfort, to satisfy human needs while using much less material and energy. You too, can be a part of this global transformation, but first, you need to learn how to deliver practical results.

Engineering disciplines

Until the middle of the 19th century, there were only two types of engineers: civilian engineers and military engineers. Civil engineers, the original civilian engineers, were responsible for most engineering projects at the time. They designed and organised the construction of railways, roads, water supply and sanitation systems, canals, ports, and buildings. The Industrial Revolution in the 18th and 19th centuries led to the proliferation of manufacturing and machines; mechanical engineers specialised in the design and manufacture of machines, as well as the rapidly growing number of consumer products. The invention of electric machines in the 19th century naturally resulted in the development of electrical engineering as an engineering discipline in its own right.

In the early years of the 20th century, the industrialisation of chemical science necessitated the development of chemical engineering as a further major engineering discipline. Software engineering emerged in the late 20th century as computer programming increasingly became a significant focus of development. Environmental engineering emerged from growing concern about the need to maintain the quality of our environment and reduce the destructive impact of human activities.

Even with about 350 engineering disciplines (also known as fields of engineering or engineering branches), at the latest count, the work performed by engineers in all disciplines is remarkably similar. From a technical point of view, practically all engineers rely on the same basic concepts in engineering science:

- Equilibrium: Things stay the same and only change in response to an external influence.
- Mathematics: Provides the underlying mathematical logic in which the laws of engineering science are framed.
- Systems Thinking: Engineers define a boundary around a cluster of interacting objects and artefacts, and then think about what must cross the boundary and what must remain inside.
- Conservation Laws: Energy, momentum, charge, and mass must be conserved. Except in the context of certain nuclear reactions, mass remains constant.
- Continuity: The notion that what goes in must come out, somewhere. It can't disappear.

Engineering students learn these fundamentals in their first 2 years of study.

What makes up the rest of an engineering discipline is a language of ideas, design methods, knowledge of relevant materials, components and systems, and an appreciation of supporting resources, practices, and ideas, along with a wealth of detailed technical knowledge.

As we shall see in the coming chapters, the part that is common in all disciplines—coordinating technical collaborative work—takes most of the time and effort in real-world engineering practice. There are only minor differences in certain disciplines.

Many, if not most, engineers end up practising in a completely different discipline from the one they graduated in. Because most engineering methods are common to all disciplines, it is not difficult for an engineer to work in several disciplines throughout their career.

Table 2.1 Engineering disciplines

No matter which discipline an engineer graduates in, there are many others that he or she can specialise in by starting with the same fundamental knowledge. This list will help you understand which others you might move into from your present focus. It is not an exhaustive list; you may have opportunities to go into a completely different discipline at some point in your career.

Multidisciplinary engineers These engineers can come from any discipline. Several marked with asterisks involve highly technical work in one or more disciplines.		
contracts engineer	inspection engineer*	project manager
cost engineer	maintenance engineer*	research and development engineer*
engineering academic*	military engineer*	scheduler
engineering asset manager*	patent attorney*	simulation engineer*
estimator	production engineer*	technical standards engineer
facilities engineer*	production supervisor*	technical writer*
heritage engineer	project engineer*	test engineer*
Chemical engineering disciplines		
biochemical engineer	oil and gas engineer	process engineer
bioengineer	operations engineer	process safety engineer
biomolecular engineer	ordnance disposal engineer	protein engineer
bioprocess engineer	paper engineer	quantity surveyor
chemical engineer	patents attorney	reaction engineer
corrosion engineer	petrochemical engineer	reactor engineer
crystal engineer	petroleum engineer	reactor operator
explosives engineer	pharmaceutical engineer	refinery manager
food processing engineer	plant engineer	refractory engineer
genetic engineer	plant operator	site remediation engineer
health and safety engineer	polymer engineer	textile engineer
mineral processing engineer	process design engineer	tissue engineer

(Continued)

Table 2.1 (*continued*) Engineering disciplines

Environmental engineering disciplines		
aquaculture engineer	geographic information system engineer	port engineer
bioengineer		river engineer
biomechanical engineer	geotechnical engineer	signal processing engineer
bioresource engineer	groundwater engineer	site remediation engineer
city engineer	hydraulics engineer	smart infrastructure engineer
ecological engineer	hydrologist	
environmental engineer	irrigation engineer	sustainability engineer
erosion engineer	natural resources engineer	tailings engineer
finite element analysis engineer		urban engineer
	ocean engineer	wastewater engineer
flood control engineer	oceanographer	water supply engineer
forest engineer	offshore engineer	wind engineer

Civil engineering disciplines		
agricultural engineer	flood control engineer	railway engineer
airport engineer	forensic engineer	remote sensing engineer
architect	foundation engineer	reservoir engineer
architectural engineer	geographic information system engineer	river engineer road engineer
biomechanical engineer		
bridge design engineer	geological engineer	sanitation engineer
builder	geomatics engineer	seismic engineer
canal engineer	geotechnical engineer	site engineer
city engineer	groundwater engineer	site remediation engineer
civil engineer	highway engineer	smart infrastructure engineer
coastal engineer	hydraulics engineer	
combat engineer	hydrologist	structural engineer
concrete engineer	infrastructure engineer	subsurface utility engineer
construction engineer	irrigation engineer	surveyor
construction management engineer	local government engineer	sustainability engineer
		tailings engineer
construction planner	maintenance planner	traffic control engineer
disaster relief engineer	mine manager	traffic engineer
drainage engineer	mining engineer	transport engineer
earthmoving engineer	municipal engineer	tunnelling engineer
earthquake design engineer	ocean engineer	urban engineer
ecological engineer	oceanographer	utility engineer
environmental engineer	offshore engineer	wastewater engineer
erosion engineer	pavement engineer	water engineer
explosives engineer	port engineer	water supply engineer
finite element analysis engineer	quantity surveyor	wind engineer
	quarrying engineer	wind tunnel engineer
fire protection engineer		

(*Continued*)

Table 2.1 (continued) Engineering disciplines

Computer engineering disciplines		
application engineer	digital systems engineer	robotics engineer
application support engineer	embedded systems	signal processing engineer
AI engineer	engineer	smart infrastructure
chief information officer	expert systems engineer	engineer
code designer	field support engineer	software architect
computer engineer	human factors engineer	software configuration
computer hardware and	information systems	engineer
systems engineer	engineer IT systems	software engineer
customer support engineer	manager	software support engineer
data communications and	mission specialist	space engineer
networks engineer	network engineer	submarine engineer
data scientist	(computer, IT)	traffic control engineer
database engineer	network engineer (IT)	web engineer
digital signal processing	operating system	
engineer	engineer	
Electrical engineering disciplines		
acceptance test engineer	hotel engineer	quality engineer
aerospace engineer	human factors engineer	quantity surveyor
airport engineer	hydropower engineer	radar engineer
analogue electronics engineer	industrial automation	radiation protection
audio engineer	engineer	engineer
avionics engineer	industrial design engineer	reactor operator
biomedical engineer	industrial process control	recording engineer
bionics engineer	system engineer	reliability engineer
broadcasting engineer	instrumentation engineer	remote sensing engineer
building services engineer	lighting engineer	renewables engineer
chip design engineer	logistician	robotics engineer
city engineer	logistics engineer	safety engineer
clean room engineer	machine condition	satellite engineer
computer engineer	monitoring engineer	semiconductor engineer
computer hardware and	mechatronic engineer	signal processing engineer
systems engineer	medical equipment	signalling engineer
configuration management	engineer	site engineer
engineer	medical imaging engineer	smart infrastructure
control engineer	medical instrument	engineer
customer support engineer	engineer	sound engineer
data communications and	microelectronic engineer	space engineer
networks engineer	mission specialist	studio engineer
digital signal processing	nanotechnology engineer	submarine engineer
engineer	network engineer	substation engineer
digital systems engineer	(electrical power)	switchgear engineer
disaster relief engineer	network engineer (IT)	systems engineer
electrical engineer	neural engineer	telecommunications
electronic engineer	operations engineer	engineer
elevator engineer	optical engineering	

(Continued)

Table 2.1 (continued) Engineering disciplines

embedded systems engineer	optoelectronics engineer	test equipment engineer
energy safety engineer	photonics engineer	traffic control engineer
field support engineer	plant engineer	transmission engineer
flight control engineer	power electronics engineer	underwater acoustics engineer
flight engineer	power generation engineer	underwater weapons engineer
forensic engineer	power systems engineer	value engineer
high voltage engineer	process engineer	video engineer
	procurement engineer	weapon systems specialist
	protection engineer	weapons engineer
		wireline logging engineer

Mechanical engineering disciplines

abattoir engineer	heating and ventilation and air-conditioning engineer	reliability modelling engineer
acceptance test engineer		renewables engineer
acoustics engineer	hotel engineer industrial design engineer	reservoir engineer power transmission engineer
aeronautical engineer	industrial engineer	powerplant engineer
aerospace engineer	industrial process control system engineer	powertrain engineer
agricultural engineer	infrastructure engineer	pressure vessel design engineer
air-conditioning engineer	instrumentation engineer	process engineer
aircraft engineer	integrity engineer	procurement engineer
aircraft structure engineer	irrigation engineer	product design engineer
airframe engineer	kinematician	propulsion engineer
apparel engineer	lighting engineer	prosthetics engineer
architectural engineer	locomotive engineer	pump engineer
astronautics engineer	logistician	quality engineer
automotive engineer	logistics engineer	quantity surveyor
ballistics engineer	lubrication engineer	radiation protection engineer
bearing engineer	machine condition monitoring engineer	railway engineer
bioengineer	machine designer	reactor operator
bioinformatics engineer	maintenance planner	refinery manager
biomechanical engineer	manufacturing engineer	refractory engineer
biomedical engineer	human factors engineer	refrigeration engineer
boiler engineer	hydraulics engineer	robotics engineer
building services engineer	hydropower engineer	rocket engineer
cellular engineer	industrial automation engineer	rotating equipment engineer
ceramics engineer	marine engineer	safety engineer
city engineer	marine surveyor	satellite engineer
clinical engineer	materials engineer	scheduler
combat engineer		
composite structures engineer		
condition monitoring engineer		

(Continued)

Table 2.1 (continued) Engineering disciplines

configuration management engineer	mechanical design engineer	seismic engineer
construction engineer	mechanical engineer	ship engineer
construction management engineer	mechatronics engineer	signal processing engineer
construction planner	medical equipment engineer	site engineer
control engineer	medical imaging engineer	smart infrastructure engineer
cryogenic engineer	medical instrument engineer	sound engineer
customer support engineer	metallurgist	space engineer
designer	mine manager	sports engineer
die design engineer	mineral processing engineer	structural engineer
disaster relief engineer	mining engineer	submarine engineer
drill support engineer	mission specialist	subsea engineer
drilling engineer	mobile equipment engineer	subsurface engineer
elevator engineer	mould design engineer	supply chain engineer
energy engineer	nanomedical engineer	systems engineer
engine design engineer	nanotechnology engineer	test equipment engineer
environmental engineer	naval architect	textile processing engineer
ergonomics engineer	nuclear engineer	thermal engineer
explosives engineer	nuclear weapons specialist	thermodynamics engineer
fabric engineer	offshore engineer	tissue engineer
fatigue specialist	oil and gas engineer	tool design engineer
field support engineer	operations engineer	traffic control engineer
finite element analysis engineer	optomechanical engineer	transmission engineer
fire protection engineer	ordnance disposal engineer	tribologist
flight control engineer	packaging engineer	turbine engineer
flight engineer	patents attorney	underwater acoustics engineer
fluid power engineer	petroleum engineer	underwater weapons engineer
food-processing engineer	pipeline engineer	value engineer
forensic engineer	piping design engineer	vehicle engineer
forging engineer	plant engineer	vibration engineer
fracture mechanics engineer	plant operator	water supply engineer
furniture design engineer	plastics engineer	weapon systems specialist
gear design engineer	pneumatics engineer	weapons engineer
geothermal engineer	port engineer	welding engineer
health and safety engineer	rehabilitation engineer	wind tunnel engineer
	reliability engineer	yacht design engineer
		yarn engineer

Further Reading

Trevelyan, J. P. (2019). *30-Second Engineering.* London: Ivy Press.

Chapter 2

Engineering practice

Engineering practice is mostly invisible.

We can see the results of engineering every day. Phones, buildings, roads, vehicles, and aircraft: the list of engineering achievements is almost endless.

Yet, these are all objects, some of them vast systems of man-made structures, while others are almost too small to see.

Engineering practice, on the other hand, is a human activity. It is performed by extraordinary people in some cases, but most are entirely ordinary. Engineering artefacts like drawings, models and documents represent what is to be, or what has been built: the finished objects. What they seldom represent is the human performances that led to these artefacts and the creation of the objects that they represent.

Engineers are the people who conceive, plan and organise the delivery, operation, sustainment and eventual removal of all these objects. Engineering practice is what the engineers do to make all that happen. Many other people play a part: people with specialised knowledge and skills, from finance to practical trades. Usually, engineers compose only a small proportion of all the people involved in an engineering enterprise, which might include hundreds or thousands of people.

Figure 2.1 A newly completed rail tunnel. Engineers are invisible in most images of engineering. Tunnels are invisible except for the entrances: people pass through in darkness, seeing little. The face of the person is invisible. Engineering practice, as explained in this chapter, has also been largely invisible. (Photo: Ricardo Gomez Angel unsplash.com)

Many engineers will tell you that they hardly do any 'real engineering': design, calculations and problem-solving learned in engineering school. Instead, their lives seem to be filled with what one described as 'random madness', seemingly trivial and routine paperwork, meetings, phone calls, frustrations, confusion, and misunderstandings.

In the words of Dilbert, the cartoon engineer created by Scott Adams, "My job involves explaining things to idiots, who make decisions based on misinterpreting what I said. Then, it is my job to fix the massive problems caused by the bad decisions."

One of the great controversies in understanding the ancient world concerns the construction of the great pyramids of Egypt. Even with the prolific hieroglyphic writing that litters the remains of the ancient Egyptian empire, no one has been able to find any records that explain how the pyramids were built. Maybe none were written at all.

Perhaps engineers today are no different from their Egyptian forebears. The documents and artefacts that we engineers create represent the endpoints of our performances. How these artefacts came into being—the human engineering performances we call engineering practice—is no more likely to be written down now than it was 4,500 years ago. It remains as it always has been: performed every day, yet simultaneously invisible.

It is hard to learn about something that is invisible. This book, therefore, will help lift Harry Potter's cloak of invisibility that perhaps he carelessly left lying over the secrets of engineering practice. The aim is to make engineering practice visible and explain what, to most engineers, has been something taken for granted. Figure 2.2 attempts to visualise some of the elements of engineering practice.

Engineering is a collaborative enterprise where specialised technical knowledge, foresight and planning are essential. Three human capabilities provide essential support:

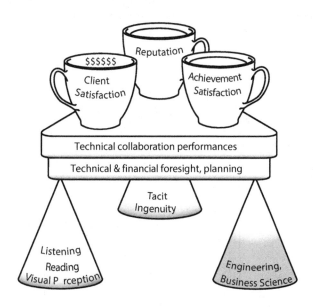

Figure 2.2 Elements of engineering practice.

i   Engineering and business science: the extent of coloured shading qualitatively indicates how much is taught in university courses.
ii Perception skills – listening, reading, and seeing: these skills, often overlooked, form the foundations for communication and collaboration abilities, and are rarely mentioned in universities.
iii Tacit knowledge and ingenuity – accumulated through experience, mostly without any awareness – is also largely overlooked in university courses but is the foundation for creativity and rapid decision-making.

The three cups represent value generation and satisfaction: the outcomes of successful practice. Client satisfaction results in payment for services, while reputation leads to future opportunities.

Value creation rests on a platform of technical and financial foresight, planning, and technical collaboration performances. None of these are taught in engineering schools, and the simple representation in the diagram hides many complexities that will be discussed later. The aim of this book, therefore, is to introduce and expose the elements of practice that receive limited attention in engineering schools today.

Another way to view engineering practice is illustrated in the following diagram, Figure 2.3. It shows engineering as an activity with two distinct threads.

The upper thread consists of three phases. It starts with discerning, comprehending, and negotiating client and societal needs, and then discussing engineering possibilities with clients and the broader society through various means. The second phase is conceiving achievable and economic engineering solutions that will meet those needs. The third phase is performance prediction: working out the technical performance and the cost of those proposed solutions.

The lower thread is all the activities needed to deliver solutions that meet the needs and requirements on time, on budget, safely, with the predicted performance, and with acceptable environmental and social impacts. Without this second thread, engineering would not yield anything useful. Solutions on paper, no matter how elegant,

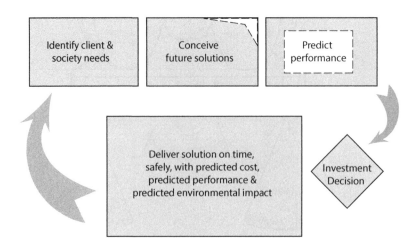

Figure 2.3 A simplified model of engineering practice.

provide little value until they can be translated into reality. Effective engineers are able to deliver; they get things done.

The large arrow represents the idea that experience in the delivery aspects of engineering in the lower thread enables an engineer to develop expertise that helps with discerning client requirements, creating new ideas, and predicting performance in the upper thread. This works like a feedback loop: engineers progressively develop more and more expertise as they learn from experience.

The small white areas enclosed in dashed lines represent the deeply technical components of engineering practice: this is what many engineers refer to as 'real engineering'.

The remaining parts still require engineers' technical expertise, but there is a stronger socio-technical element required for effective technical collaboration. These much more substantial aspects dominate the time and attention of engineers. That's why the core of this book is devoted to technical collaboration. Unfortunately, these dominant aspects of engineering practice are hardly mentioned in engineering schools, which is why many engineers find it hard to label much of their work as 'real engineering'.

How to use this book

This book introduces the fundamental principles of engineering practice in 20 short chapters.

Learning is hard work: it takes much more than just reading the text. Engineers are expected to learn all the time. Therefore, adopting a systematic approach to learning will help you learn with less effort.

We all learn faster if we are more aware of what we are learning.

The Professional Engineering Capability Framework document provides a curriculum for learning engineering practice. It can be accessed at the online supplement for this book, also at https://www.jamesptrevelyan.com/. Additional material for this book can be found at the book page on the publisher's website: https://www.routledge.com/9780367651817.

Since the 1990s, it has been customary to describe the attributes required for engineering practice as 'competencies', concisely described combinations of knowledge, skills, and attitudes. However, a detailed awareness of engineering practice is needed to appropriately interpret competency definitions.

For example, communication is often cited as an essential competency for engineers. Universities expect students to demonstrate their communication abilities by preparing PowerPoint presentations and reports: this is how they interpret 'communication' for engineers. However, arranging for concrete to be delivered to the right place on a construction site at the right time requires completely different communication skills.

Therefore, the Professional Engineering Capability Framework lists 'performances' that a supervisor could expect from you in the first 3 years of practice. You need to learn these performances to practice independently as an engineer, without supervision. Some of the performances are simple. Others require complex social interactions: you will need practice and feedback to reach a professional standard of performance.

I recommend that you discuss a few of these performances each time you meet with your supervisor during face-to-face meetings, ideally once a week, as they become relevant. Accumulating the evidence of these performances and recording when you achieved them at a professional standard will serve as a record of your learning. Your work achievements documented by these and other records can help you apply for Chartered Engineer status with professional engineering organisations in your country.

Most of your records will be noted in your work diary, preferably a notebook with blank pages. It contains ideas, sketches, notes of meetings, phone calls, the times you worked on different tasks, and even records of casual interactions with other people. It's an essential record of daily work as an engineer. Your work diary could also be electronic, on your phone, and voice recognition has reached the stage where it's beginning to be a convenient option. It's not yet so easy to combine freehand sketching and electronic notes. I use Dragon for professional voice recognition because it can handle the specialised vocabulary of words and abbreviations. I still use a physical notebook as my main diary, but I am acutely aware that it is easily misplaced.

Just one note of caution. Increasingly, smart devices are being restricted in workplaces for security reasons. Never lose the ability or instinct to use pencils and paper.

Keeping detailed records of your work is essential for many reasons. For example, you may need it to justify invoices for consulting work. One day, you could be required to defend your decisions in court, and your diary could provide crucial written evidence that you exercised appropriate professional due diligence many years earlier.

Read a few pages of this book each week. Ideally, you should set aside time in your calendar (see Chapter 19 on time management). Plan a regular weekly meeting with yourself to do this.

Towards the end of each day, write a series of dot points in your work diary describing the work performed, meetings attended, important phone calls, and other significant events not already recorded.

On the last working day of the week, review the book pages you read during the week and write out your own summary: a set of dot points summarising the ideas introduced in those pages.

Then:

a   Write a set of dot points summarising what you learned during the week that was related to the ideas you've read in the book so far.

b   Write a few paragraphs, no more than about 400 words, about one of the learning experiences you encountered during the week that helped you see how ideas in the book play out in practice.

In some chapters, there are additional exercises: these should be completed as well.

Chapter 3

Seeking paid engineering work

Seeking a paid engineering job is the necessary first step in an engineering career. It can be an emotionally draining time. It certainly was for me: it took months, and I was depressed and irritable. It would have been much less frustrating had I known about this job-seeking method.

Even if you don't get a job immediately, by following the instructions in this chapter, you will be much more employable as an engineer after a few months. If you don't get a job within 6 months, please write about your experiences and send a copy of your job-seeking diary so that I can use your feedback to improve the next edition. So far, after 25 years of providing this advice, I have yet to hear about a single instance when this method has not worked.

Here are two common misconceptions about seeking engineering work.

1. Engineering jobs are always advertised
2. You cannot get an engineering job without relevant experience

Eighty to ninety percent of jobs are not advertised: companies rely on suitable people coming forward when they need them. This is called the informal job market. The only way to find these jobs is through networking: going out to meet other engineers and people who work with engineers, as explained in this chapter. Building your network makes you more employable as an engineer.

Companies nearly always prefer to hire an experienced engineer. However, if they cannot find someone with appropriate experience, they will then look at other applicants, a decision that might require the provision of appropriate training.

Applying for advertised jobs is often time-consuming and can be a waste of effort. Even if you make the shortlist, you can expect to spend days on tests and interviews.

Fear of failure

For most engineering graduates, the fear of a negative reaction is the greatest obstacle that holds them back. If you feel the same way, you're not alone. Courage is taking action despite the fear: remember that we all share the same fears. You will only succeed if you try.

Stage 1: Preparation

Step 1: Create your job-seeking diary, build your job-seeking contact list

Obtain a notebook (size A4 or A5) and write your name and contact details on the front page to make sure it can be returned if you misplace it. Use this notebook to record everything you do in terms of seeking work: all the names of the people you meet, their contact details, and everything you see or hear that is related to your search. Create an electronic version on your phone. Ensure that you have a secure back-up process in case you lose your phone.

Note all new contacts and record when and where you met; document every subsequent interaction with them with brief notes.

Step 2: Start building your network of contacts

Ask everyone among your friends and family if they know anyone with connections to engineering projects. An uncle or aunt may not be an engineer, but they may be able to introduce you to one who can be your starting point.

Keep in contact with your class colleagues, as many of them will be able to suggest contacts for you.

Join the organisation for engineers in your discipline. In Australia, there is a single organisation 'Engineers Australia' for all engineers. Other countries may have separate organisations for different disciplines. Join as a student or graduate member: most charge only nominal fees. Go to local meetings. At each meeting, make a point of getting to know one or two engineers. Note their contact details and ask them to suggest people you can visit. You will find that most engineers are very helpful, as they have all been through the same experience as you.

Ask people about conferences and trade shows in your areas of interest: attending can help you find many more contacts.

You will continue building your network of contacts long after you secure your first job. Your contacts list is one of the most valuable documents you will create in your career.

Step 3: Prepare your résumé and online profiles

You must have an up-to-date résumé as a written document that you update from time to time, along with an online profile—LinkedIn is preferred. You should also create profiles on recruitment sites, such as Seek.com.

Written communication skills are important. Companies will employ an engineer with less than perfect spoken English if they think they are technically competent and able to communicate effectively. However, a poor résumé or profile will generally result in your application not even being considered. Arrange for someone to proofread your résumé and profile.

A résumé provided for prospective employers should be short, two pages maximum, and can refer to your LinkedIn profile or another website for more details.

There are many good books on preparing a résumé. There are suggestions at the end of the chapter that should be available through your local library or online bookstore.

While books can provide you with a lot of helpful guidance, you should seek feedback on your résumé, preferably from an engineer with experience in hiring or a careers adviser. These books also provide valuable tips on interview skills.

Common mistakes

Most companies reject 75%–95% of the applications they receive because of one of the following reasons:

- The cover letter was either missing or did not mention the needs of the particular company.
- The résumé was poorly formatted or contained obvious spelling errors.
- The descriptions of the work that the applicant had previously performed did not give a clear idea of their capabilities. For example, the applicant might have stated, "I was involved in the installation of solar photovoltaic systems." Instead of this, describe the actual work you were required to do, such as,

> I had to liaise with clients and installation contractors to arrange the best time to install solar photovoltaic systems. I had to confirm these arrangements just prior to installation to make sure that both clients and contractors were ready and prepared for the installation work with appropriately trained people, tools, equipment and materials.

Describe the actual work that you performed, even if it does not seem to involve what you think of as technical engineering work.

Review your online presence

When a company is interested in employing you, they will almost certainly review your online presence on sites like Facebook and LinkedIn for potentially compromising or embarrassing material. Review your entire online presence and delete any material that could be embarrassing for your employer, whether it is in English or any other language! If possible, ask a responsible friend to search on your behalf as well.

Key attributes

The following attributes stand out in a résumé and increase your chances of success.

LEADERSHIP

An engineer always has to be a leader, even in a purely technical role, without management responsibilities. An engineer who appreciates what needs to happen and then inspires other people to collaborate and help achieve it is valuable in any organisation. You can demonstrate your capacity for leadership in various roles, such as organising activities in clubs or community groups, particularly fundraising. You can also demonstrate your capacity for leadership through your part-time work record, such as acting as a shift supervisor in a restaurant.

TEAMWORK

Playing a team sport or working on an organising committee helps to demonstrate teamwork abilities, not necessarily in a leadership role.

INITIATIVE

You display initiative when you initiate actions without waiting for instructions or requests from other people. Demonstrate that you have researched engineering organisations and have managed to get yourself past the company reception and speak with technical staff.

In many cultures, demonstrating initiative can seem unnatural, as though one is stepping out of line with accepted social practice. In these cultures, it is common for younger people to remain silent and respectful in the presence of older, more experienced people. Acting with initiative in these cultures requires different methods from the more open, less hierarchical European, Australian, or American cultures.

PERSISTENCE

Persistence is highly valued by engineering employers. You can demonstrate persistence and organisation with your approach to job-seeking. If you have or have had significant disadvantages, such as a family breakup or a major illness episode, briefly explain how you overcame them.

RELIABILITY AND RESPONSIBILITY

Describe circumstances in which other people have placed their trust in you. For example, if you have acted as treasurer for a club or society, and the members have trusted you with their money, that is a sign that you are both respected and trustworthy.

LOCAL WORK EXPERIENCE

While you are searching for a job, working part-time with any organisation can provide you with useful experience, especially if it allows you to gain some background in sales, marketing, administration, occupational health, or the safety and supervision of staff.

ABILITY TO LEARN FROM EXPERIENCE

Explain how you have been able to learn from experience and your surroundings. Even if you are working in a fast-food restaurant, there are many opportunities to learn how to handle incoming supplies, orders in progress, finance records, staff performance records, and many other aspects of running a business.

Step 4: Expand your engineering knowledge: research suppliers

One of the most valuable areas of knowledge for any practising engineer is the awareness of which firms supply which products. Review their websites, note their locations, and record contact details.

Think about the products and services that you would need to make use of in an engineering job in which you are likely to find yourself. Search for information about these products and services. Accumulate your own library of technical data sheets and service specifications, costs, delivery times, and applicable standards.

Step 5: Expanding your knowledge and skills

Make use of the time before you start work to build your knowledge and skills: this proactive work will significantly improve your employability.

Read Chapters 4–8, assess your listening, reading and seeing skills, and practice these skills to improve your performance.

Identify items in the Professional Engineering Capability Framework that you can learn and practice before you start work. These include Section 3 (intellectual property knowledge is not covered well in most engineering courses), Section 4 (occupational health and safety regulations, attend first aid course), Section 5 (practice reading, learn the history of major engineering projects), Section 7 (employment regulations), Section 11 (critical thinking), and Section 12 (Read historical accounts of major engineering projects).

Read current news on business and finance to learn about developments in major engineering-related business activities in your region. Keep up to date with developments to identify areas where investment funding is likely to create employment opportunities for you.

Observe everything that has been built around you, systematically taking notes. Although you can focus on the engineering discipline in which you have the most interest, it is important that you also observe products and infrastructure created by other engineering disciplines. You also need to observe the engineered services provided in your area, including the water supply, drainage, and sanitation, electric power, roads, buildings, communications, and transport.

Take photographs of representative installations, clearly showing their state of repair, particularly if there are obvious faults. Also, make a point of preparing sketches—from real life if possible, otherwise from photographs. Sketches can illustrate technical ideas much more clearly because you can eliminate visual clutter that cannot be removed from photographs.

Evaluate how well existing products and infrastructure are meeting the needs of people. To do this, you need to ask the people who actually make use of them; it is not sufficient to rely on your own opinion. Keep your discussions informal and sociable; ask about spare time interests. Once they are more relaxed, ask them what they think about the value of the services and products they use. Keep notes about what people say in response to your questions.

Prepare detailed records of any infrastructure that clearly needs to be repaired, particularly roads, drains, water supplies, and electricity connections. Even if it is working satisfactorily, note the condition of the infrastructure. Whenever you find maintenance work being performed, ask permission from the maintenance workers to watch and learn about the work they're undertaking. Engage them in casual conversation; you will be surprised at how much you can learn.

Standards

Build your working knowledge of relevant standards. Ask engineers in your network what standards are considered the most important in the industry in which you would like to work.

Workplace safety standards and local employment laws and regulations are some of the most important guidelines for engineers to understand. These are usually available from government websites.

If you are still a student, and your university provides access to online standards documents, make a point of downloading your own copies before you finish your studies. Some self-destructing PDF documents can be preserved by disabling JavaScript in your PDF reader. Otherwise, print your own copies and keep them as printed documents or convert them to a scanned PDF.

If you no longer have access to online standards, ask other engineers. Most will be happy to share at least the important parts of relevant standards to help you improve your knowledge.

Programming

Teach yourself how to use macro programming for software like Microsoft Office, Visual Basic, and Acrobat. Word, Excel, and Outlook all support macro programming that enables you to automate many tasks and prepare intelligent document templates. Many engineering organisations make extensive use of macros in routine engineering documents to ensure that engineers deal with all the issues required for a particular task.

Another useful programming environment to master is 'Python,' which enables you to make many computerised tasks fully automatic. All the software you need is readily available at a modest cost. You can find extensive self-study materials online.

Contractors

Through local contacts, news media, and directories such as the 'Yellow Pages', learn about small contracting firms that you would need to perform various kinds of support work needed in engineering. This includes earthmoving and excavation, cable laying, security systems, fencing, lighting, and transport. Learn about material suppliers and tool suppliers. Visit firms to establish personal contacts. Either obtain business cards and note the details, or write down names and telephone numbers and, if possible, email addresses.

Material, labour, and component costs

Visit a hardware store and learn about the costs of all the components and materials needed to construct a house. Develop a comprehensive list of all the items and calculate the cost of the components and materials needed.

Find some building contractors and learn about the cost of the labour needed to install these components and materials, and don't forget about the cost of hiring someone to supervise the work. This takes time and effort, but you will find the exercise to be valuable practice for your engineering career.

Logistics

Learn about logistics by working in a bakery or fast food restaurant. Find out how much of each material the enterprise needs to purchase in advance, how it is stored, how long it can be stored, how much is used, and how much has to be discarded because it is no longer fit to be used. Learn how the manager figures out what needs to be done and when so that the bread or cooked food is ready precisely when customers arrive.

Economics

Learn about the economics of the business. Learn about materials and labour costs. Include all the components of labour costs (see Chapter 16): not just what each person is paid but also the taxes, the cost of the building and space they need to work in, the cost of having someone to supervise, manage and train people, the cost of someone to clean up waste material, and the cost of insurance, administration, and accounting, as well as other indirect costs. Learn what the customers pay for finished products and services.

Predictions

Watch construction or other engineering-related activities that you can see in your neighbourhood. There may be trains on a nearby railway or a port with ships being loaded or unloaded. Learn to make accurate forecasts. For example, estimate:

i   time and distance needed for a train to come to a stop at a station (or a bus at a bus stop),
ii time needed for everything that happens during a stop, and the time needed to accelerate to operating speed, and
iii total space needed.

Learning to make relatively accurate predictions quickly, without having to rely on detailed calculations, helps you further develop abilities that you will need as an engineer. This is all about learning to be observant.

Watch for and anticipate breakdowns. It might be a toilet cistern in the home or a loose piece of roofing that will blow away the next time strong winds come. It might be overhead electrical wires or poles that have corroded or a noisy wheel bearing on a car.

While you are waiting for the failure to occur, calculate the likely cost of the failure. Describe the disruption that is likely to be caused by the failure. Anticipate the time needed to get repairs organised and performed. Anticipate what other activities will have to stop or be rescheduled or relocated while the repairs are being performed. Consider several possible times at which the failure might occur, for example, at the worst possible or least inconvenient times, and see what difference this makes to the cost.

Stage 2: Visit engineering suppliers and potential employers

The most productive job-seeking activity is to visit engineering component and service providers, as well as small- and medium-sized companies that employ engineers. Only visit larger companies if you have already made contact with an engineer working there and can arrange to visit that person at their worksite. (Professional Engineering Capability Framework, Sections 13c, 13d).

Talk with sales engineers. Their job requires them to have contacts on all the engineering businesses that buy the products and services they represent. Mostly they are keen to help because they know that, one day, you will be working as an engineer and you will be more likely to buy their products and services.

Step 6: Planning

Call or write a letter asking to visit suppliers to learn about the products and services they provide. Make sure you clearly state what times you would be available to meet them. Unless you already have a contact at the company, address the letter to the chief executive. Include a request that they inform you about any trade shows where the company exhibits its services and products.

If you don't hear from the firm within 2 weeks (longer over holiday breaks), telephone the company to ask when you can expect a reply and ask if they could suggest a time for a visit.

Follow up with telephone calls each week until you receive a response. Make sure you record in your job-seeking diary the name of the person to whom you speak in each case, the date and time, and whether there is a direct telephone number you can call to reach that person again without having to go through the company switchboard. Add each person you speak with to your contacts list.

Avoid following up by email unless the company has replied by email, and only then after trying to contact the relevant person by phone.

Step 7: Visiting engineering suppliers

Make sure you are appropriately dressed and take your diary.

Use the following self-introduction when you arrive at reception:

> I will soon be working as an engineer and I've come here to learn about your products and services. Before I start work, I have more time to learn about engineering products and the companies that supply them in this region. I've read your website, but I need more information than you have shown there.

Do not start by asking for, or about, employment.

The reason for this is simple. Asking about employment gives the staff a chance to say something such as "Sorry, we only advertise jobs online – we don't answer queries about jobs here." Asking for information about their products and services makes it much harder to send you away. Only companies that you would not want to work for will send away someone who may soon be buying their products and services.

Ask to speak with someone on their engineering staff or a technical sales representative because you need to discuss technical issues. Be patient; be prepared to make an appointment to return on another day. If you do make an appointment, call early on the scheduled day to confirm that the appointment time is unchanged.

Ask about the background experience of technical staff, the range of products they hold in stock locally or in the same country, stock that can be made available on-site within 24 hours, and delivery times for items not held in stock, including the time needed for customs clearance. Ask about any training courses that the supplier provides: ask if you can sit in on any upcoming training courses. Ask for materials that you can study at home and, if possible, a quick look at some actual product samples.

The engineer or technical sales representative will probably ask what kind of work you expect to be doing. Describe the kind of position you would ideally like to get but let them know that you are still undecided. If the technical sales representative doesn't ask you about the work you will be doing, then ask "Who do you think I should be talking to about getting engineering work in areas like...?"

You could even ask them for alternative suggestions for interesting places to work.

Add all the people you meet to your contacts list with full details—all this information will be useful in the future.

Listen carefully, take notes, and ask for clarification if you're not sure that you have understood something correctly (see Chapter 5). Learn as much as you can about the companies you visit and prepare detailed notes after your visit with a list of their products and services.

If the company staff do not treat you well, make a point of recording those details. In a year or two, when you're working as an engineer, if that company's staff approach you to consider buying their products or services, you might be able to use this to your advantage. At some point in the conversation when it comes to discussing commercial terms, such as prices, for example:

> There's one thing I should say that might help in this discussion. Your people probably would not remember me, but I remember my last visit to your company. Please think of making a really attractive offer to replace the memories I have had since then and create an amazingly helpful impression of your company!

Step 8: Continue researching new job opportunities

Along with visiting suppliers, you should also continue background research on potential job opportunities.

Read online or print newspapers and check online job advertisements such as 'Highly experienced engineering or mining project manager to set up a team of—'. Think about approaching companies that advertise senior positions as a prospective junior member of their new team. Wait for 4–5 weeks and try visiting the company, following the suggestions in the next section. By then, the new senior manager has probably started work, and you should ask to meet them.

Follow business news media for reports about companies that have announced plans for expansions or special projects. Then think about the types of smaller companies that may be involved in those projects and consider approaching them. Ask them what kinds of work they are hoping to win in the near future. Always find out as much as possible about each company before making an approach about working there.

Many industries have free magazines (or email/web newsletters) that advertise products used by engineers in that field. One example is 'What's New in Process Technology'. Make sure you subscribe to these magazines and read about the products in your field.

Step 9: Visiting a prospective employer

A visit to a prospective employer requires some additional preparation. The way you approach a company makes a big difference to your chances of obtaining work with them. It is always best if you have a contact within the company as a result of prior networking or through visiting engineering suppliers. However, 'cold calling' can also be an effective way to find work.

If your marks are not good (i.e., you have failed some courses), avoid larger companies and advertised positions.

Before you approach the company, think about your objectives:

a   Career? Long term or short term?
b   Vacation or professional internship employment – is money important?
c   Engineering experience? In a particular field?
d   General work experience? Would you be prepared to work unpaid for a limited time?
e   Part-time employment to help gain experience and financial support?

Learn as much as you can about a company before approaching them. Read their website, especially any content about recent annual or project reports.

Start with a similar self-introduction but explain that you are also visiting to ask about working for them.

> I've learned quite a bit about your company and have come here to learn a little more about your products and services. I'm also interested in finding a way that I could work with you to help you build your business.

Make sure you briefly describe the skills you can offer that are relevant to the company's operations, show that you have done your research on the company, and demonstrate some good reasons for them to take you seriously.

When the conversation reaches the point where they are interested in you working with them, be specific about your objectives where they seem to align with company interests.

Step 10: Follow-up opportunities and consider starting your own business

As you visit firms and develop your professional network, make sure to regularly contact all the people that could possibly help you find work. Follow up with people you meet after 3–4 weeks and ask them to update you with recent developments. This helps to keep the memory of meeting you refreshed—opportunities will have often appeared in the time since, opportunities that you will never hear about unless you stay in touch with them.

What if, despite your best efforts, you still can't find work after several months?

All the preparation work has provided you with much of the knowledge you need to start your own business. Read Chapter 14 in *The Making of an Expert Engineer* for some suggestions that can work, even in the most difficult economic times and places.

Relocating for opportunities?

It is easier to follow the advice in this chapter if you're located in a major city with many engineering-related companies. However, there are also many good reasons for pursuing jobs far from home.

First, companies outside major cities usually find it harder to attract good engineers, so you may have a better chance of gaining work with these firms. They may also offer higher pay. Firms in major cities will seek help from local experts, sometimes even for relatively minor issues. Firms further away from cities rely much more on their own engineers, so you will have greater responsibilities and opportunities to demonstrate your abilities. Of course, there are fewer social opportunities, but you may actually find it easier to develop lasting friendships in a small community. Finally, with negligible commuting, you will have far more time for recreation and possibly many unique places to visit.

References and Further Reading

Anderson, T. A. (2019). *Engineer Your Career: A Complete Guide to Landing a Job in Engineering* (1st ed.). USA: Thomas A Anderson.

Fasano, A. (2015). *Engineer Your Own Success: 7 Key Elements to Creating an Extraordinary Engineering Career, Updated and Expanded* (2nd ed.). Wiley-IEEE.

Slocum, S. L. (2018). *She Engineers: Outsmart Bias, Unlock your Potential, and Create the Engineering Career of Your Dreams.* Engineers Rising LLC.

Trevelyan, J. P. (2014). *The Making of an Expert Engineer.* London: CRC Press/Balkema – Taylor & Francis.

Chapter 4

Neglected perception skills

Communication skills are widely regarded as being the most important competency for engineers. They are the foundation skills that enable collaboration and influence in an enterprise, the means by which engineers' ideas are transformed into reality.

Most educators and short courses, however, only focus on one-way communication: writing and giving a confident technical presentation or sales pitch. The focus in the next three chapters is on neglected perception skills: listening, reading, and seeing.

These skills are vital for two reasons.

First, the transmission aspects of communication—writing, speaking, and drawing (or the use of graphic images and other visual content)—all depend on your ability to read, listen, and see accurately. For example, before you speak, you need to listen to your audience to understand the words and phrases that are meaningful for them. Accountants speak the same language—English—yet the way they use words is very different compared with engineers.

It's also important to be able to sense when your audience is not listening—continuing to speak is then a waste of time.

The second and even more compelling reason is that becoming an expert engineer is a long-term learning journey: it will take up to 10 years or more before you can count on being reasonably expert in your field of engineering. The rate at which you can acquire engineering expertise is almost entirely dependent on these three perception skills: listening, seeing, and reading. Your ability to learn depends entirely on your ability to perceive reality, and this can be much more difficult than it might appear.

Perception skills are demonstrated by accurately perceiving the world around you and noticing important details. In the past, your teachers pointed out what was important. From now on, most of the time, you will have to rely on your own senses.

The most informative conversations with other people come without notes, and you typically won't be able to record what they say. Often, people don't want to write down this kind of information at all. Even if you can make a recording, you probably won't have time to go back and listen to it for a second time. You need to quickly and accurately understand what is being said, often with heavily accented, grammatically incorrect English with speakers using words in strange and unexpected ways.

Visual skills are needed in order to notice subtle features in artefacts (e.g., cracks, crazing, distortion, and slight defects) and unusual features in graphical information (e.g., unusual data trends on a display). The ability to read drawings, which demonstrates a capacity for spatial imagination and visualisation abilities, is also critically important. Expert engineers also know that sketches, graphics, and drawings are useful ways to help convey intent and meaning to others, but the ability to prepare sketches depends on seeing accurately.

When you go on-site, you will be expected to notice what's significant. You will almost certainly have a phone camera with you, but you need to know what to point the camera at and how to take photographs that show what's important.

Many times, you will receive verbal or email reports about faults, mistakes, and failures. Almost invariably, these reports will only provide a partial description. Whenever possible, visit the people involved, and inspect the site yourself; listen carefully to everyone who can provide helpful information. If you can, offer to help people responsible for rectification work, even if it means getting your hands dirty. You will learn far more this way than from the reports you receive at your desk.

In other words, you need to reacquire the ability to use your own ears and eyes. That takes time and practice.

For you to become a competent and articulate engineer, capable of influencing your engineering enterprise, these three perception skills will be essential. Recent results from psychology research can help you understand just how difficult it is to master high-level perception skills.

Before we get into details, however, it is important to understand that these skills are just the beginning, the foundational skills required for collaborating effectively with other people. In later chapters, we will focus on these collaboration skills, building on the foundation of perception skills in the next four chapters.

Perceiving reality

Seeing what's really there requires us to understand how prior knowledge can both help us and deceive us. The world is often different from what we expect. Learning only really occurs when we discover that our expectations don't match reality.

It is very difficult to make sense of the world without some prior belief, an expectation about what we are experiencing. Yet, these expectations and beliefs can also prevent us from seeing reality, which is why learning is very difficult. To learn effectively, we need to understand how our normal senses are influenced by prior beliefs, and then temporarily 'disable' that influence.

The images in Figure 4.1 illustrate the power of prior knowledge. If you have not seen these images before, they can be hard to discern. Once you know what they portray, it's hard to see them as before... relatively meaningless patterns of black and white. If you have trouble seeing what they portray, check the end of this chapter, then look again at these images and note how differently you see them afterwards.

For many years, cognitive scientists held that perception was a layered process that started with a low-level analysis of visual scenes or sounds. For example,

Figure 4.1 Illustrating the power of prior knowledge.

sharp changes in brightness often indicate the edges of an object. Perceiving edges, colours, corners, and other features was presumed to be the most fundamental step in visual perception. Everything else, it was thought, depended on identifying these details first.

Now we know that human perception is much more complex. The ways we reliably make sense of what we see and hear depend on prior knowledge about what we expect to see and hear. Our beliefs shape our perception, yet they can also deceive us, as shown in Figure 4.2. An interesting TED talk by Beau Lotto[1] provides many entertaining examples that demonstrate this relationship.

In later chapters, when we take up the challenges involved in influencing other people, remember this discussion of prior knowledge and how it can influence perception. This can help you appreciate how difficult it is for other people to learn something new, so that you can begin to influence what they do. It is vital that you work through the practice exercises. They are an essential foundation for the rest of this book.

1 https://youtu.be/mf5otGNbkuc

Figure 4.2 Demonstration of the 'Cornsweet Illusion,' in which the flat surface of the upper grey tile appears darker than the shaded surface of the lower white tile, whereas in fact, they are the same shade of grey, as shown in the following squares sampled from the image above. Covering the white and dark shading between the tiles with a finger or pen will help to expose this truth. Our beliefs about the tile colours influence our lower-level perception of the shades of grey (Purves, Shimpi, & Lotto, 1999). Image reproduced by permission.

Image samples taken from the flat surface of the upper tile (left) and the lower tile (right).

Prior knowledge influences perception

We depend on prior knowledge for interpreting drawings and photographs, just as much as we do to interpret speech and words on paper. Sometimes this can lead us astray. For example, a mechanical engineer once explained to me how a large and very expensive reactor vessel made from exotic material was specified with drawings using the third-angle projections. The third-angle projection is a drawing convention for describing three-dimensional objects using two-dimensional views (projections) of the object from the top (plan view) and sides (elevation views). While the third-angle convention is common in the Americas, the mirror image first-angle convention is commonly used in Britain and former British dominions. In this instance, the reactor vessel was fabricated by technicians who mistakenly interpreted the drawings using the first-angle convention, even though the third-angle symbol was included on all the drawings. They made a mirror image of the specified reaction vessel, and by the time the mistake was discovered, it was too late.

The next four chapters focus on listening, reading documents, reading people, and seeing, respectively. Remember that all perception skills depend on prior knowledge. Learning to distinguish reality is a skill that has to be acquired by practice.

Note for Figure 4.1: The black-and-white image on the left shows a silhouette of Napoleon Bonaparte wearing his characteristic hat on the left, between what appears to be two trees. The silhouette is loosely based on the portrait painted by Francois Pascal Simon Baron Gerard. On the right side, the image shows a black-and-white Dalmatian dog, nose to the ground at the centre of the image, with its rear left side towards us.

Reference

Purves, D., Shimpi, A., & Lotto, R. B. (1999). An empirical explanation of the Cornsweet effect. *The Journal of Neuroscience*, 19(19), 8542–8551. doi:0270-6474/99/198542-10$05.00/0

Chapter 5

Listening

This is a brief introduction to learning accurate listening: the most important communication skill.

Engineers spend between 20% and 25% of their time listening, on average; that's far more than any other single activity they perform.

Accurate listening and note-taking do not come naturally. These are acquired skills that need to be learned and practised. It is hard work, just like physical fitness training. Improving your listening skill is one way to improve your 'emotional intelligence', a term used by psychologists to assess your ability to collaborate with other people.

There are many engineering performances that completely depend on accurate listening; below are two examples.

First, gaining finance: nothing is possible in engineering without money, usually lots of it. Most clients with money tend to be verbal people: they don't typically express their needs in writing or drawings. That's the first reason why listening is so important for engineers: you need to listen in order to thoroughly understand your clients' needs. Your clients will usually only approve funding for your work once they are confident that you truly understand their needs.

Second, collaborating with others. Engineers cannot achieve much without a lot of help from other people. You need to be sure that they are listening to you, and you need to listen to them. Accurate communication is one of the best ways to avoid nasty engineering problems. The major contributing factors to most engineering disasters have been communication failures. The Transocean crew operating the Deepwater Horizon rig drilling the BP Macondo well in the Gulf of Mexico tried to communicate their concerns about the condition of the drilling operation to their superiors, but the risks of a catastrophe were never fully appreciated. Eleven members of that crew died in the subsequent fire and explosions.[1] Knowing when someone else has properly listened and understood is as important as listening itself.

Most people think that hearing is the same as listening. However, just a few minutes of observation will tell you that those two skills are very different. Many people can benefit from improving their listening skills, including you.

---

1 It's ironic that BP chose the name from the novel by Nobel Prize winner Gabriel Garcia Mendez. In the novel Macondo is the name of a town cursed with eternal bad fortune.

When I asked my students about which aspect of communication skills they would most like to improve, their most common response went like this:

> I would like to be able to get my point across more often. I find that other people don't listen to my ideas. I'd like them to listen more carefully because I get frustrated when they seem to misunderstand what I am saying or miss the point completely.

Knowing more about listening skills can really help you in these situations.

The chances are that the 'other people' are not listening particularly well, but you haven't noticed. Also, you probably haven't listened carefully enough to them to realise that they have, after all, understood more about what you were telling them than you thought. Communication is a two-way street, requiring the effort and attention of all parties involved.

If you can recognise the level of other people's listening skills and notice when they're not listening, you can save yourself a lot of trouble. Once you can tell when another person is not listening, you should simply stop talking. Continuing to talk is a waste of your time. Work out how to regain their attention so the information you're offering can be effectively understood.

Practice exercise: observing listening lapses

Join a group of people talking about something: it could be a project meeting, a casual conversation, or just a group of people trying to organise a social activity. Even better, if you're living with two or more other people at home, simply observe a routine conversation around the dinner table.

Watch and listen carefully.

See if you can notice when someone starts speaking before another person has finished talking.

When this happens, the interrupting person switched their mental focus a few seconds before they opened their mouth to figure out what they were going to say. They will have missed what the other person said in the last few seconds before that.

Now, think of three other indicators that could alert you to a listening failure.

1. _____
2. _____
3. _____

(See the online appendix on listening skills if you need suggestions.)

Repeat this exercise whenever you can.

The main trick for good listening is to keep your focus on what the other person is saying, right through to the end.

It's hard, and sometimes tiring, until you have practised this skill and it becomes natural.

If you're like me, you will often find your attention drifting. How many times have you been listening to someone, perhaps in a meeting, and found yourself thinking about something completely irrelevant?

Additional material can be found on https://www.routledge.com/9780367651817.

Active listening and paraphrasing

Active listening is a special type of interactive conversation, one in which the listener carefully uses occasional participation to help the speaker along. This is a skill that you can easily master, and it can make listening more fun and enjoyable. It is also very useful in meetings: your active listening will help other people in the meeting understand what someone is saying and promote additional engagement.

There are several good texts to help you learn active listening, an important skill. I recommend *People Skills* by Robert Bolton. A useful text for emotionally charged situations is *Crucial Conversations* by Patterson, Grenny, McMillan, and Switzler.

One of the key skills to learn is paraphrasing. It takes practice and can feel embarrassing at first. However, it sends a powerful message to the speaker that you are really paying attention and genuinely respect what they are telling you. When they see your level of interest, the speaker will probably tell you much more than they would have otherwise.

After a speaker has said something that's important to accurately understand, ask them to listen to your own interpretation of what they just said and tell you if it's right:

If I heard all of that correctly, what you just told me is ____ ____ ____ ____ ___.
Is that what you meant?
   or
Did I hear you correctly when you said ____ ____ ____ ____ _____ ____ _____?

This is particularly helpful in meetings, especially if you are the meeting chair. If you are unsure whether you really understood what the speaker was saying, the chances are good that the other people present have possibly also misunderstood what was said. By asking for clarification, or better yet, paraphrasing what you think the speaker just said, you will help other people understand more accurately.

You might think this takes extra time and trouble, or you might feel that you will annoy the speaker by doing this. However, you will actually make the speaker feel more reassured that you are sincerely trying to understand what they have shared.

Writing accurate notes

Being able to write comprehensive notes that capture the essence of what was said is a vital aspect of listening skills.

One common method used by journalists is to write in shorthand.

Another technique that I have used since I was a student is based on 'mind-mapping.' When I first started taking notes, I used to write key phrases in more or less the same sequence as my lecturers, something like this:

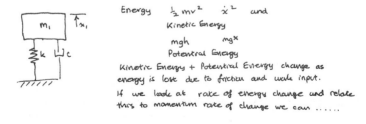

Instead, try this:

As you listen to the speaker, write down the main ideas as single words or brief phrases on the page.

It doesn't matter where you start writing, but the middle of the page is often a good choice.

Leave a little space between each of the phrases or words. As the speaker continues, write more keywords or phrases near those that are related, and draw lines or arrows showing the relationships. As you notice connections that the speaker has not referred to, you can add more arrows and connections, even adding a note or two of your own alongside the lines to explain your mental associations.

You may end up with what seems like a disorganised maze of words and lines all over the page, but the initial layout is not important.

As soon as possible after the speaker has finished, add extra notes to clarify ideas and connections from memory. Highlight important ideas or comments. I enclose any thoughts or side observations of my own in square brackets to distinguish them from what the speaker said.

When you need to recall what was explained, start with the highlighted notes. Then, follow the arrows and lines to recall what the speaker talked about; your memory will fill in the gaps between the words on the paper. After a little practice, you may be surprised that you can remember far more than the words actually written on the page, sometimes even in a more organised way than the original speech.

The diagram below required about 30 words, some lines, and some symbols to represent around 350 words of text. Given that most people speak at around 100 words per minute, you only need to write approximately 10 words per minute to keep up with what they're saying and still be able to recall everything important.

Contextual listening

The third aspect of listening requires you to develop sensitivity to particular word choices by the speaker. Let's start with a simple example of a quotation by an engineer:

They stuffed that one up really badly; it took weeks to recover the drill stem they had dropped down the hole.

Suppose this engineer had started with the word 'we' instead of 'they'. Would that have made any difference?

Either way, the drill stem—possibly thousands of metres of steel piping with an expensive diamond drilling bit on the end—has been dropped down a deep well. The choice of words would not alter the situation, but the choice of words can tell us how the speaker views responsibility for the accident.

By using the word 'they', the engineer has implied that he or she is not associated with the people responsible for the incident. Using 'we' would have conveyed the idea that the engineer is part of the group responsible and is therefore sharing some of the responsibility.

Observations like this enable you to perceive important social relationships. Understanding relationships can help you understand more about the meaning of the words people use, and perhaps the words that they choose not to say, as well as how they might behave in the future. This understanding can also help you ask relevant questions and, with practice and care, avoid unnecessarily causing offence.

Helping others to listen

You can apply your understanding of listening skills to help others listen more accurately when you are speaking.

For example, eye contact helps immensely. If I look into the eyes of listeners, shifting from one person to another every second or two, I can hold their attention longer. I can also detect when I start to lose them, as some people might start to move their eyes around the room instead of looking at me. Soon after that initial sign, if I don't regain their attention, I will start to hear shuffling feet and papers being moved around. At that point, I know that I've lost their attention.

Some people feel uncomfortable being stared at, particularly in a small group or a one-on-one situation, and will actually look away while you are talking, even though they're still listening carefully. Other people may have a natural squint: their eyes seem to be looking somewhere else, even though they're actually looking straight at you.

When it comes to lectures and presentations, remember that a PowerPoint presentation is a great attention diverter. People will look at the screen and may soon stop listening to you entirely. It is difficult for many people to read and listen at the same time. With more than ten words on the screen, people who naturally prefer reading to listening will have shifted their attention and likely stopped listening. Among engineers, most will read the text and stop listening to the speaker.

Sometimes, a picture can tell the whole story, with minimal commentary. If this is the case, stop speaking for a few moments so your listeners can devote their attention to the picture. When you want to regain your audience's attention, simply press the '.' key: the screen will go blank. (Press it again to get the picture back.) Alternatively, you can insert a black slide into your presentation, which is a clear signal for the audience to shift their focus from the picture back to listening.

An imperfect, interactive, interpretation performance

Listening is not a one-way process that starts with hearing and ends with making sense of what we hear. Instead, prior knowledge is needed to accomplish effective listening perception: we cannot make sense of what we hear without some prior understanding of language, for example. However, prior knowledge can also interfere with listening. Partly to try to resolve this potential weakness and partly to evaluate our listening performance, we engage in conversation with a speaker, asking questions to help resolve apparent ambiguity or misunderstandings. Therefore, listening is a truly interactive performance; not a one-way input process.

In some cultures, asking questions, especially with a speaker of higher social status, can cause offence. In these situations, find out from other people how to engage in an extended conversation with the speaker, possibly in private. Sometimes, it will only be possible to resolve ambiguities by discussing what was said with others.

Perfect listening is rarely possible in real situations. Even with the best interaction, the closest relationship, extended conversation, and explicit clarification, misunderstandings persist as a result of the prior knowledge that we use to make sense of the spoken words. Because of this, listening is always an act of reinterpretation: the listener is reconstructing the ideas being explained by the speaker. That reconstruction can never be the same as the speaker's original ideas.

The challenge for engineers, therefore, is to learn how to ensure that this reinterpretation by the listener still allows technical ideas to be faithfully used or reproduced, even though there will always be a certain degree of misunderstanding.

We shall learn in the next few chapters that reading and seeing are also imperfect, interactive, interpretation performances.

More listening and note-taking exercises

1   At any meeting, or even during a casual conversation, ask permission to take notes. After taking notes for 10–15 minutes, reconstruct what was said briefly as bullet points. Ask the speaker(s) to review your bullet points and check for mistakes, misunderstandings, or significant omissions.

2   Download the listening skills worksheet from the online appendix. Use this to observe different listening behaviours while watching other people in conversation or sitting in a meeting. Make sure you do this discreetly, or ask permission first, as many people can be quite offended if they think you're trying to watch them too closely.

3   The Australian Broadcasting Corporation (ABC) Radio National website provides a valuable resource to practise listening and note-taking. The website provides podcasts of many programmes, as well as full transcripts. (http://www.abc.net.au/radionational/)

  i   Each day, listen to any recorded programme that interests you, for which ABC also provides a transcript. Listen to no more than 7–10 minutes while taking notes. Do not press pause: let the recording play at normal speed.

  ii   After completing your notes, reconstruct what was said as best you can from your notes: a set of bullet points is sufficient.

   iii Use a word processor to open the transcript from the ABC website, or print out a copy of the section of the transcript corresponding to the part of the podcast that you listened to.

   iv Highlight all the text in the relevant section of the transcript that was accurately conveyed by your bullet points.

   v Do not highlight any important words or ideas that you missed, or where a word you wrote does not correspond to the word in the transcript. For example, if you wrote 'specification' (or abbreviated to spec'n), but the word in the transcript is 'requirements', do not highlight the transcript, even though the words can have similar meanings.

   vi Estimate the percentage of the transcript text that you have highlighted to calculate your score.

Many of my students have started with a score of 5%–10%. However, after practising a few times, they have improved to 50% or better. Others who started at a level of about 50% have improved to about 80%, some even as high as 90%. Individuals interpret the evaluation criteria differently, so comparison with others is not meaningful. However, anyone can monitor their own progress using this technique.

Having started on what will normally be a lifelong journey to improve your listening skills, it is now time to turn to reading. Beyond listening, engineers spend almost as much time reading, and, like listening, accurate reading is surprisingly difficult to master.

Additional material can be found on https://www.routledge.com/9780367651817.

References and Further Reading

Bolton, R. (1986). *People Skills*. New York: Touchstone Books.

Patterson, K., Grenny, J., McMillan, R., & Switzler, A. (2012). *Crucial Conversations* (2nd ed.). New York: McGraw-Hill.

Trevelyan, J. P. (2014). *The Making of an Expert Engineer*. London: CRC Press/Balkema - Taylor & Francis, Chapter 6.

Reading documents

A comprehensive reader is a person who can engage with drawings or text in detail and see different meanings and interpretations.

Only a minority of people are comfortable with reading extended texts. Your ability to acquire information from documents may mark you as a valuable information source. Therefore, it's wise to develop the skill of reading comprehensively so that you don't lead others astray.

In essence, reading is no different from the other principal modes of perception: seeing and listening. Comprehensive reading depends not only on perceiving the actual words that are written on the page but also on the prior beliefs that we have to begin with. Our perception depends on these prior beliefs, but at the same time, they can deceive us.

Comprehensive reading does require that you read every word. When we read silently, we often skip words; sometimes, we bypass entire sentences or paragraphs without being aware that we missed part of the text.

One of the best ways to make sure that you read every word on the page is to practise reading out loud to the wall of your room: imagine that you have an interested listener with you in the room. This is also valuable practice for writing: reading out loud for 15–30 minutes a day helps to build tacit knowledge in your head about the patterns in which words regularly occur. Eventually, this will make it easier for you to write better. You could read this book out loud. However, reading novels written by outstanding authors will help you write better.

Just as with listening, try to be mindful of your mental focus while you're reading. If your mind starts to wander as you read and you can't remember the content of the last few paragraphs or sentences, stop reading. Take a break and rest. Reading accuracy is significantly affected by fatigue, both mental and physical.

One way to assess the accuracy of your reading is to attempt the practice exercises in this chapter.

You can also find a wide variety of helpful texts on how to read quickly (speed reading). This technique depends on the conventional way of writing in Western countries: the main idea in each paragraph will be conveyed by the first sentence. Therefore, you can read quickly by skipping from paragraph to paragraph; you don't have to read the rest because you know that it only elaborates on the idea conveyed in the first sentence. Speed-reading can be a great way to build up a map or a framework of the content of the book and its main ideas so that you can return to the parts that are most informative for you and read them carefully at a later time.

The ideas that you acquire through reading may soon fade in your memory unless you take some action to reinforce those memories. There are several ways to do this.

Take notes as you read. These days, I use the iAnnotate App on my iPad whenever I can. Kindle has some similar capabilities. I can mark up the text by highlighting or underlining. Later, iAnnotate can extract those sections of the text into a summary document for me, giving me the page number where they are located in the original PDF document. When reading technical papers, I note the location of a quotation on the page: 35a.4 means page 35, left-hand column (a), 40% of the way down from the top of the page.

Another way to reinforce and explore the ideas that you have just read is to imagine that you have another person sitting in the room beside you. Explain the sense of the text you have just read so that they could understand the content. You may also have certain questions in your mind. Ask the imaginary person, "I don't quite get this part, when it says (X). What do you make of that?" You can even imagine what the other person might say in response, such as "Why don't you look it up on the Internet and see what Wikipedia says about it?" Can you remember a time when something that happened resembled the situation described in the text? Explain to your imaginary companion what actually happened and explain to them how reading the text has helped you understand that episode differently.

An additional way to reinforce your memory is to write questions about parts of the text you don't fully understand. Write some questions that you could ask other people around you in order to explore these ideas.

Here is one way to assess your reading when there are no practice exercises provided.

Choose any paragraph at random after you have finished reading a chapter. Reread the first sentence of the paragraph. Using your notes and memory, write a summary of the rest of the paragraph as accurately as possible. Then, read the rest of the paragraph and compare your summary with the actual text.

I recommend that your supervisor or mentor also read this book. By doing so, they will be able to help you understand more about their engineering experiences. Inevitably, they will understand the text differently from you. Therefore, by discussing the content of each chapter with them, you will learn different ways to interpret the same text.

Practice exercise: reading documents to learn from them

Like active listening, comprehensive reading is a two-way conversation. However, the original author is no longer present, so the conversation must be with oneself or another person. Here is a practice exercise that illustrates some of the ways to conduct such a conversation in order to help with learning.

Read a section of a technical report or published article between 5 and 20 pages long.

Write a brief description of the content, including the main ideas presented in the text (beyond what is already given in a summary or abstract, if available).

Write your assessment of the reliability and strength of the evidence presented to support the ideas in the text. Sometimes, there is none. Try to distinguish between parts of the text based on systematically collected evidence and other parts that resemble 'recipes' based on personal opinion.

Note any references or further reading that you think you need to follow up on. Often, we may be tempted to classify opinions without evidence as 'subjective' and unreliable. Yet, we often find that opinions can offer insights that more systematic research may miss. A comprehensive reader can make sense of diverse contributions, reconstructing ideas that may not have even been apparent to the writer of the text.

Next, explain what the writing contributes towards the questions and issues relevant to your own aims that motivated you to read the text in the first place. It can be very helpful to quote sentences or even full paragraphs from the text, but you must explain how each quote is particularly meaningful for your goals and interests.

Include a precise page reference for each of these comments.

Finally, write down your own impressions and issues arising from the text, as well as questions that remain in your mind that can stimulate an ongoing conversation with yourself and a later search for meaning.

If I were able to converse with you, the reader, as you work your way through this book, we would both learn a great deal. As I cannot be with the vast majority of you, I suggest that you write a reflective journal as you read this book, section by section. The following practice exercise will give you some pointers.

Ultimately, the test of your reading of this book lies in one or more of the following:

a   insights that you develop into the world of engineering in which you immerse yourself every day;
b   a demonstrable performance improvement that brings personal rewards for you, the reader, in terms of job satisfaction and (preferably) improved remuneration in the long term; or,
c   the accumulation of evidence to demonstrate that the ideas in this book need revision or restating, in which case I hope that you will send your accumulated evidence to me so I can take it into account for a subsequent edition.

Practice exercise: written requirements

Even though many clients find it difficult to provide engineers with written requirements, there are some that do provide detailed written statements of requirements. Furthermore, engineering work often has to be performed so that it complies with contract conditions, recommendations contained in standards, or explicit requirements stated in codes. Engineers need to carefully read these documents to ensure that they allow time for any work that will be needed to meet the requirements and demonstrate compliance. In these circumstances, comprehensive reading is essential.

The following exercise will help you develop comprehensive reading skills.

If you have not tried this before, you may be quite surprised to find how much you miss when reading a complex document. Some engineering companies formalise this process; in those companies, it is normal practice for at least two people to read a statement of requirements and exchange notes on what they found to reduce the chance that something important has been missed.

Two or more people read the same text and prepare notes. While reading the document, each reader must note down each of the requirement clauses that call for some action or response and give each requirement clause a priority rating score, as shown below:

4 = critical: cannot be missed;
3 = important: compliance will enhance value;
2 = necessary, but not important: compliance will provide minor additional value; and
1 = optional: compliance will not influence the value of results.

When both readers have finished their notes and have listed requirement clauses with priority ratings, compile a list of all the clauses noticed by both readers. Then calculate the following:

N = total number of clauses identified by both readers
Ma, Mb = number of clauses identified by a reader (a or b) as a proportion of N

Identify clauses where both readers assigned the same priority rating.
   Calculate an interpretation score for each reader Pa, Pb: Add all the priority ratings for clauses where both readers agreed. Divide this score by the sum of all the priority ratings assigned by the respective reader to obtain his or her interpretation score.
   Here is a small worked example, with the scores given to a series of clauses in the following table. Shading shows where both readers identified the same priority rating.

Clause #	Priority given by A	Priority given by B
1	4	4
2	1	2
3		
4	2	2
5		3
6	3	
7	3	3
8	4	
9	3	4

N = 8 (Total number of clauses for which a response was seen as necessary by at least one reader)
Ma = 7/8 = 0.88 (Proportion of clauses identified by reader A as needing a response)
Mb = 6/8 = 0.75 (Proportion of clauses identified by reader B as needing a response)
Pa = (4 + 2 + 3) / (4 + 1 + 2 + 3 + 3 + 4 + 3) = 0.45

(Proportion of clauses identified by reader A with agreement on priority)

Pb = (4 + 2 + 3) / (4 + 2 + 2 + 3 + 3 + 4) = 0.5

(Proportion of clauses identified by reader B with agreement on priority)
   The closer your M and P values are to 1, the more consistent your reading is in comparison to the other reader. However, a score of 1, apparent perfect consistency, may indicate that you have both missed something important. To be sure, prepare a

summary of the critical requirement clauses and ask someone more experienced to see if anything important has been missed.

Words and natural language can never be understood in the same way by every reader. Words are mere symbols, or sounds, associated with ideas in our minds. We have all had different educational and life experiences, and we have different memories, so one's ideas associated with any one word are always different from another person's ideas. That is why everyone interprets the same text differently: it is part of being human. So, arranging for two or more people to read a single set of requirements is more likely to expose different meanings and interpretations. These interpretation differences must either be clarified with the client or allowed for when planning the work needed to fulfil the requirements.

Does this seem like hard work?

Like any aspect of expertise, developing comprehensive reading skills takes practice, which requires effort. It takes time and persistence and can be much more fun if you have the support of friends and colleagues who can help you on your journey.

While reading documents can be challenging, reading people can be even more difficult. The next chapter takes you on a brief tour before we deal with seeing.

Chapter 7

Reading people

Why is there so much emphasis on listening?

Engineering is an influence game. As engineers, we don't build anything, and rarely if ever, work with our hands, other than to type on a keyboard or hold a phone. We can normally only achieve something through the hands of other people, using someone else's money. An experienced maintenance engineer explained it like this: "No amount of engineering calculation, drawing or writing ever changes anything until a fitter does something different with his tools." Even if you end up inventing, designing, making, and distributing a gadget or software app entirely by yourself, success depends on influencing people to buy it and gain satisfaction from using it.

Influencing the behaviour of other people requires that we understand what motivates human behaviours.

Emotions are the most powerful influence on behaviour. That's why it is so helpful to learn to 'read people'; to observe their emotions.

Psychologists use the term 'emotional intelligence' to describe these attributes and abilities and many firms will use psychometric tests to evaluate them as part of their recruitment process.

Unfortunately, many engineers have a tough time when it comes to emotions. Through all our years of education, emotions are hardly mentioned at all. We learn that emotions have no place in rational discussion; logic should prevail over 'mere human emotions'. You might hear someone say, "Leave your emotions at the door—they have no place here at work." As a result, many engineers find it hard to read emotions in other people, and perhaps more so within themselves.

We need to understand that emotions, while mostly invisible, are seldom rational. One of my graduate students expressed it like this: "People don't obey Newton's laws. When pushed in one direction, they move in a completely different direction."

Careful observation is the only way to read emotions, and the easiest way to do this is to engage a person in a conversation, face to face. With some experience, especially if you already know a person, this can be done with a phone call, or even better with a video call. However, face-to-face encounters provide the best setting, which is why effective engineers spend so much time talking face to face with other people. It is by far the most effective influencing technique.

Mostly we can read emotions through non-verbal cues, often called body language. Most of us can sense the resistance expressed by the girl in the photo. There are many internet sites that provide images and videos easily found by searching for 'body

Figure 7.1 Example of body language.

language'. Most of these resources are specific to American culture; other cultures have similar 'body languages', but there are also significant differences (Figure 7.1).

In this short chapter, therefore, I will explain some of the ways that emotions influence social behaviour in some common engineering scenarios. The best example is territory.

Land, or territory, is important for all of us. Someone entering 'our' space—our territory, especially our home—is labelled as 'an intruder'. In extreme cases, we talk about an 'invasion' when unwelcome people enter our territory.

Think about Graham, an electrical engineer responsible for designing all the cable arrangements and connections required for a project. He can reasonably expect that he would be consulted on any changes affecting electrical cabling, connections, the spaces allocated for them, and the equipment that the cables connect.

If someone else makes changes that affect cable arrangements, especially if Graham was not consulted, we might expect a strong reaction. It is easy to think that a small change, a change that might make it easier to access or install the cables, would be welcomed, and no consultation would be needed. After all, that kind of change can only help. And changes like this often occur in engineering projects.

Graham might not react at all . . . immediately. However, it is safe to assume that this intrusion into 'his territory' will create an emotional response. He may not show any visible signs. He might even forget about it. However, normally there would be some resentment that—when combined with something else, perhaps completely unrelated or perhaps combined with many other small resentments—suddenly causes a long-delayed reaction. The reaction might even be hard to associate with the original territorial intrusion. It could take the form of an entirely logical objection on some other issue, something that to others seems inconsequential. However, Graham's resentment may have motivated him to build an argument fortified with extensive detailed evidence, and no one else knows enough to see the emotional scars through the argument.

Gender issues can also intrude into emotional discussions. It is common to think that women are more capable with human relationships, can be more emotional than men, or that they can handle emotions better. Others might think that even discussing emotions is inappropriate for men, that talking about emotional issues is a female attribute. Try substituting a woman's name for Graham in the preceding paragraphs. Would you see the discussion differently?

Another example of territory comes from research observations of photocopier maintainers. Photocopier machines that were regularly serviced by particular technicians, and were seen to be the responsibility of those technicians, were 'their machines'; part of their 'territory'. These machines were significantly more reliable than other machines. The other machines, maintained by technicians who were randomly assigned on a given day, tended to fail more often. One of the explanations offered was that a technician who knew he or she would be coming again would notice small signs indicating future failures, perhaps some tiny scraps of torn paper under the paper feeder or a stain from leaking lubricant. They would know that they could easily perform some preventative maintenance that would save them from having to carry out an arduous repair task on a later visit. Pride in workmanship is easier to maintain with a sense of personal responsibility and that leads to higher quality work and greater machine reliability.

Engineers who see a particular area of technical, or even non-technical, work as their 'territory' may take more care and exercise a greater level of responsibility for 'their' piece of the project.

Avoid email and text messages for sensitive conversations

Email and text messages seldom resolve conflicts; usually, they make them worse. Novices often think that email and text messages are fast and easy. Additionally, you don't have to face an angry response to bad news.

Expert engineers know that when there is any conflict or emotional arousal, there is a good chance that any message will be interpreted in a way that further inflames the feelings of conflict. Any emotional arousal, whether positive (e.g., affection, happiness, excitement) or negative (e.g., anger, frustration, fear), is likely to increase the chance that the message will be interpreted in a way that reinforces the perception of that emotion. Furthermore, many people who are angry or frustrated want to distribute their response far and wide in the organisation, partly as a means (they think) to garner support and sympathy from others. They may do that by sending an email reply with lots of people copied in on the message.

When talking face to face, with the ability to read body language, it's possible to pause or even defer a conversation in response to the emotions of the listener. One can even stop and ask the listener to explain their feelings, and then decide whether to address the emotional issue first, before continuing with the original conversation.

It's harder to do this in a phone call. Even in a video call, which only shows the face of the listener, it may be hard to perceive any hand movements betraying the emotional state of the listener.

When using text, this kind of immediate feedback is absent. The reader, with perhaps heightened emotions, tends to interpret the text in a way that reinforces their feelings, which is why disputes can escalate rapidly when people only communicate in writing (Figure 7.2).

Figure 7.2 An engineer inspecting an amusement park ride to ensure critical safety interlocks is functioning correctly. The amusement park manager behind her is watching anxiously, as his business could be closed down for expensive repairs if she cannot certify that the ride is safe to operate. She needs to be aware of the emotional pressure that his presence may be creating to ensure that her report provides an objective safety assessment: lives could depend on her judgement.

From time to time, we all face situations where we have to deliver bad news to someone. These can be very difficult experiences, so be prepared to seek help and support from others. Above all, consider the person receiving the message and make sure that they have adequate support to deal with the message when they receive it.

Further reading

Boyatzis, R. E. (2009). Competencies as a behavioral approach to emotional intelligence. Journal of Management Development, 28(9), 749–770. doi:10.1108/02621710910987647

Chapter 8

Seeing and creativity

Being able to see is crucial in engineering, especially for learning.

There are visually impaired engineers, of course. However, they rely on others to see for them. This chapter will help you evaluate your ability to see. If you think that your seeing skills could improve, an online supplement to the book provides a graded set of exercises that you can work on independently over several weeks.

Additional material can be found on https://www.routledge.com/9780367651817.

The best way to assess your visual perception skills is freehand sketching, possibly with annotations. It is only when we sketch what we have seen that we can be sure that something has been noticed, in the same way that a paraphrased response to a speaker can help confirm that we have listened to and understood the intentions of the speaker (Figure 8.1).

As engineers, most of us see ourselves, in some way, as creative designers, or at least as creative people that can innovate and devise original engineering solutions that meet the needs of our clients.

Creativity is the ability to come up with original and innovative ideas when needed. Good ideas don't simply emerge from nothing. Creativity depends on accumulating a vast memory of ideas and observations in your mind; that internal library is tacit knowledge of which you are normally never even aware. You can't write it down,

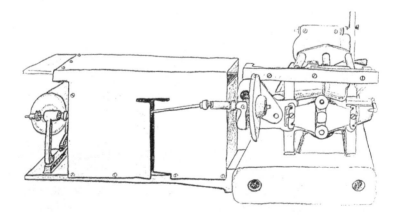

Figure 8.1 Sketch of a tensile testing machine by a student.

since you often can't remember that you possess it. However, when you need it, these memories emerge . . . but not always when you want them, of course.

Design expertise, in particular, relies on a vast memory of design ideas. Every time you see an engineering artefact, whether it is a culvert under a road, a bridge, a machine, an optical waveguide, a special connector, or any one of hundreds of other artefacts, you can potentially store this in your mind. However, this only happens if you learn to see the details, which depends on accurate seeing skills. The best way to accumulate these memories is to sketch them, freehand, at every opportunity in your work diary.

Except for engineers with significantly impaired vision, most learn more with their eyes than any other sense, but it is a mistake to think that just because you can see with your eyes, you can see accurately (remember, just because you can hear doesn't mean you are listening). As we will soon see, your eyes and brain, when working together, can easily deceive you.

Why is sketching so difficult?

Can you write your name? Can you draw reasonably straight lines on a piece of paper? Can you draw a square or a circle? Most people can do these simple tasks reasonably well. However, drawing the person sitting next to you or even your own hand can seem like an impossible challenge.

The reason for this stark increase in difficulty is simple. You can write your name or draw a square because your brain is more than able to control the movement of the pencil on a piece of paper. The only reason why drawing the person sitting next to you seems impossibly difficult is because your brain substitutes the image perceived by your eyes with a preconceived idea of how to draw a person. This preconceived idea takes the place of what your eyes actually see. Accurate seeing, therefore, requires that we master the mental discipline needed to suppress this automatic response by our brain. Learning to see means learning how to stop your brain from taking over your mind, successfully putting aside preconceived ideas and allowing your pencil to reflect what your eyes are actually seeing, just like a photocopier.

'Seeing' is an active process that you can improve with training and practice. It is a state of mind that requires discipline and the ability to put aside distractions and preconceived ideas. The quality of our sketching, therefore, reflects our ability to see.

Learning to draw accurately is like learning a sport: you need to build your tacit knowledge, which is the knowledge that connects your eyes with your fingers and enables you to accurately reproduce what your eyes see.

Training is essential, as well as practice. Remember that practice can be very tiring, so build up your stamina in stages, starting with no more than 30 minutes of training each day.

Practice exercise: evaluate your seeing skills

You may be surprised when you attempt these exercises. If you have learnt technical drawing skills or the use of CAD software like AutoCAD, Pro-Engineer, or SolidWorks, you may think these exercises will be easy for you. Many of my students who had good technical drawing skills have been surprised by how difficult they actually are.

For this evaluation exercise, you will need

- Several A3 sheets of paper and also a cardboard base (size 42 × 30 cm)
- A clutch pencil (0.5 mm, 2B lead) or 2B wood pencil
- Eraser

If you have a digitiser tablet or touch screen and a sketching app or basic version of Photoshop, you can do the following exercises without using paper or pencils.

Reduce distractions as much as possible. You need a quiet room with good diffuse lighting, preferably free from direct sunlight and sharp-edged shadows. You also need a comfortable chair. Classical music may help you concentrate—works by Mozart and J. S. Bach are often recommended.

Before you start, take a few minutes to relax and let go of the tension and distractions of the day that have built up thus far.

You will do four evaluation drawings. Take no more than 30 minutes for each and be sure to write your initials and the date on each of them.

Drawing 1:

- Draw a square, a circle, and a rectangle, side by side.

Drawing 2

- Draw a person sitting in front of you: either the whole figure or just a head. If you're in class, draw a fellow student. If you are alone, use a mirror to draw a self-portrait. If you can't get access to a live figure, draw a portrait by working from a photograph of a person. Write the date and your initials when you have finished.

Drawing 3

- Draw your own hand.
  Place your left hand (your right hand if you are left-handed) on one side of an A3 sheet of paper and prepare a drawing of your hand on the opposite side of the paper.
  If you think that you might be interrupted, lightly trace around just the tips of your fingers before starting so that you can return your hand to the same position on the paper.

Drawing 4

- Place a shiny metal spoon on a book and draw them together.

Take a short break. Make sure each drawing has your initials and date.

Now, judge the results. You are probably much more satisfied with your square, rectangle, and circle than you are with your portrait or the drawing of your hand.

You managed to write your name? What this shows is that your ability to move the pencil is not an issue. The only reason why the other drawings were not as good is that your eye and brain combination is not yet allowing you to move the pencil in appropriate ways.

How well do the shape and shading of your portrait correspond to the likeness of the person? Would you recognise the sketch as your hand? Can you see the reflections in the shiny metal parts of your drawing of the spoon? Can you see the texture of the binding of the book?

The shortcomings in your sketches demonstrate that your ability to see can always improve. When you can see accurately, you will be able to draw accurately too.

Like listening skills, almost certainly, your ability to see can be improved, and one of the best ways is to learn freehand sketching. Books such as Betty Edwards' *Drawing on the Right Side of the Brain* provide ample advice with lots of enjoyable practice exercises, many of which can be done while at work.

Further Reading

Edwards, B. (2012a). *Drawing on the Artist Within*. New York: Touchstone.

Edwards, B. (2012b). *Drawing on the Right Side of the Brain* (4th ed.). New York, Tarcher.

Gadd, K., & Goddard C. (2019). *TRIZ for Engineers: Enabling Inventive Problem Solving*. Hoboken: Wiley.

Trevelyan, J. P. (2014). *The Making of an Expert Engineer*. London: CRC Press/Balkema - Taylor & Francis, Chapter 6.

Part 2

Workplace learning

The first part of this book helps you prepare for an engineering career. You can work through the chapters while job-seeking. Even if you have already started at work, these early chapters are essential preparation for the second part of the book. The second part helps you learn engineering practice from your workplace experiences.

The book does this in three different ways.

First, the book points out important aspects of engineering practice and provides you with words to describe them—a vocabulary of engineering practice. For example, Chapter 10 explains different kinds of knowledge used by engineers, and Chapter 11 explains how this knowledge is distributed among the people you encounter in the course of your work.

Next, each chapter explains important concepts that can help you understand the patterns of human behaviour in an engineering workplace. These are called performances: actions and social interactions that engineers follow as part of their work, particularly to collaborate with other people.

Specific performances that you will be expected to demonstrate are listed in the Professional Engineering Capability Framework document provided in the online appendix on https://www.routledge.com/9780367651817.

Some performances are relatively complex, involving many distinct stages of interaction with other people. Therefore, it is helpful to explain them at a higher level, as in Chapter 12 on technical coordination, a kind of informal leadership, and Chapter 14 on project management.

Third, I expect you to keep records of your work in your diary. I also expect you to adopt a weekly routine, reading a few pages, even a complete chapter, and discussing your performances with your supervisor. Meeting less often with your mentor is also helpful.

Your learning will be more effective if you follow the recommendation to write brief reflective notes at the end of each week, describing what you have observed in the workplace in terms of ideas explained in the book.

Chapter 9

Learning the ropes

> The first priority for a young seaman in the days of the great sailing ships was to learn the names of the hundreds of different ropes used to raise, lower and control the sails, even to steer the vessel. Hence the term 'learning the ropes', describing the first few days and weeks of a life at sea. Of course, there were so many other things to learn as well.

You have an engineering job and your first day has arrived, at last. You have practised your listening and note-taking skills in the last few days, and you have your notebook with you to start your work diary.

You may have experienced a roller-coaster ride of emotions, from intense anticipation and excitement on receiving the phone call confirming your job, to the fear of failure. "Will I make a fool of myself?"

Almost certainly, the first week will be completely different from your expectations.

If you start with a large corporation, the first week (or more) may consist of formal inductions, a welcome from a senior manager, endless forms to be completed, registration processes to access email and information systems, and detailed workplace policy documents and safe working regulations to read and understand. You may soon find yourself on an extended series of training courses on company processes, health and safety, project management, documentation control systems, and many other topics. Few of these courses will have a technical agenda, much to the frustration of many graduate engineers.

In a small company, your experience can be much less predictable. You may be expected to be contributing to a project team on the first day. You may have a supervisor who has shown you around to meet everyone, has a desk ready for you, and a set of briefing documents for you to read. Equally, you may have a supervisor that forgot you were coming that day, promises to lead you around, but is tied up in meetings all week. You may be asked to analyse data to diagnose a performance issue in a process you barely understand. You may even find yourself in a role where you are expected to be guiding technicians, even engineers, with decades of experience.

As a novice, you will be starting on a 'steep learning curve'. Expect to spend far more time learning than doing for the first few months. However, it can be difficult to know what you need to learn first.

The Professional Engineering Capability Framework will help (available in the online appendix on https://www.routledge.com/9780367651817). It can guide you through the first years of your career as you 'learn the ropes'.

It lists engineering performances that a supervisor could expect from you in the first 3 years of practice, framed in terms of Australian engineering competencies that are compatible with and recognised by all the Washington Accord member states. The evidence that you have demonstrated these performances at a professional standard will help you apply for chartered engineering status in any country.

The competencies were written with the help of the research that led to this book. Being able to demonstrate these performances helps confirm that you are ready to practise engineering, unsupervised, as an engineer in your own right.

Understanding the principles of engineering practice explained in later chapters will help as well. These chapters provide detailed guidance on aspects of practice—for example, what you need to know as an engineer, how to gain the willing and conscientious collaboration of colleagues and others, and how to work safely (Table 9.1).

There are 16 sections, one for each competency. The start of each section defines and briefly elaborates on each engineering competency.

Table 9.1 Sample section from the Professional Capability Framework

I. Personal commitment: Dealing with ethical issues	
Means that you anticipate the consequences of your intended action or inaction and understand how the consequences are managed collectively by your organisation, project, or team; and means you demonstrate an ability to identify ethical issues when they arise and to act appropriately.	
Year 1	*Date notes location ref.*
a. Describe situations that more experienced people have encountered when ethical issues, dishonesty, or conflicts of interest were apparent to them.	
b. Describe restrictions on distribution of information and explain the reasons for restrictions.	
c. Explain the consequences for people who do not have access to information or foresight to know what might happen to them.	
d. With help from supervisor, identify stakeholders and their interests, explain how those interests are affected by the consequences of engineering actions. Include, as stakeholders, people who are not easily able to be represented in discussions, including future generations (Section 5 lists examples of stakeholders).	
Years 2–3, in addition...	
(aa) Describe workplace ethical standards, explain why personal integrity and honesty are important for individuals and the organisation.	
(bb) Describe situations in which information must be shared with other people, and explain how stakeholder interests are advanced by sharing this information.	

A list of performances follows that your supervisor can expect you to learn during your first year with the firm. After those are additional performances for years two and three, but you may master some in your first year. Take the sequencing only as a suggestion.

Practice exercises (not shown above) will help you improve. Discuss these with your supervisor or mentor—they may offer many other useful suggestions. Informative learning resources are listed below them.

The right-hand column provides space to record the date(s) when you first achieved a professional level of performance, judged by your supervisor, and also to record where the evidence (notes) has been stored.

The most important person you will meet when you start at work is your immediate supervisor. He or she will be your guide. Most supervisors look forward to the time when you will be independent and need minimal support. So, listen carefully, take notes, and learn as quickly as you can. Make sure you have their contact details, at least their email and mobile phone number.

You should also look out for one or more mentors who, ideally, should be outside your firm. You may be lucky and have an engineer as a family member. If not, ask at your local professional engineering society office, or attend some meetings and ask engineers that you meet there. Your mentor can give you a different perspective to help you develop your career with a broader focus, beyond the bounds of a single company.

Of course, if your immediate supervisor is not an engineer, finding an engineering mentor will be even more important. It can take time to find one; in the meantime, continue building your professional network.

Start at Section 2 of the Professional Capability Framework. It describes performances needed for competent practice. You will find guidance on working with your supervisor:

(2a) Discuss and understand needs and requirements with client or supervisor, write notes, ask clarifying questions, interpret in terms of engineering possibilities to seek confirmation, write summary, review the summary face to face with client, or supervisor to confirm.
(2b) Write clear and concise agreement, summary, or specifications of work requirements and completion criteria, keep track of changes as requirements and completion criteria evolve.

Your technical work, of course, will be determined by your discipline and the type of work that your firm is performing. You may have the choice of office work or practical experience on-site or in a factory. Even though it can be demanding and involve unpleasant locations, always take opportunities for site work when you can.

It is not unusual for much of your early career work to involve measurement and performance evaluation, described by the following performance:

(16a) Analyse data relating to the performance of a product, process or system, and compare results with predicted or specified performance. Measurements or simulation data can be used, depending on context and requirements.

Equally, you could find yourself on-site, helping with construction work or manufacturing.

You will need information about the product, project, process, or system that you will be working with. This is classified as Local Knowledge in Section 13. Here are some relevant performances you will need to master.

(13a) Operate information systems to access data, specifications, project plans, CAD documents, codebase, and software defining product, process, or system, including associated tooling, fixtures, all other relevant software, documentation, and data. Can access configuration documents and systems that define the current and historical status of product, process or system, and variants.
(13b) Operate information systems for correspondence (email, messaging transcripts), document storage and archiving, human resource management, payroll, purchasing authorisation, and expense claims.

Engineering information nearly always changes with time as people introduce modifications and improvements. Thus, it's equally important to know where to find configuration information that tells you how product and process information has changed at each revision. Often you will need to know, for example, the details of a process as it was at some time in the past—its current configuration may be quite different (Figure 9.1).

If your supervisor has planned for your arrival, you will be provided with documents containing the information you need to get started. However, you will soon need to go beyond the initial documents. You will probably need information that is not even available in company documents.

In university, you learned to find information for yourself, often using internet search tools. Hopefully, you also learned how to locate more reliable sources. However, most

Figure 9.1 Three fundamental information resources: your laptop to access information resources, your work diary or notebook, and a printer for documents used for marking up and discussions. Some people prefer electronic markup, but printed documents are often more comfortable for round table discussions.

of these were only accessible through the university's paid subscriptions to databases, such as journals and indexed sources like the ASME, ASCE, ASHRAE, INSPEC, Engineering Village, Knovel, and Xplore digital libraries, as well as collections of engineering standards. You may no longer have access to these sources. There are some public domain sources, such as Engineering Toolbox and Wikipedia, but discretion is needed when using them—always seek authoritative sources as a check on their reliability.

Even in well-organised companies, accessing all the data you need can be a complex process in itself. You may need special software tools on your company laptop or mobile devices. Almost certainly, your information technology (IT) support people will need to configure information systems to allow you to access the information. In some firms, drawings and paper reports may still have current information, so you will need to know where to find them and whom to ask for permission to access them. You may require an appropriate level of security clearance to access the most sensitive information.

In most firms, emails and notes of telephone conversations, records of meetings, etc. are vital records that people need and commonly refer back to. You will need to learn how these records are stored and preserved.

Many small firms have no formal systems or procedures for this. You may be surprised how much goes missing when a company laptop is either lost, stolen, or 'retires' with its owner when they transfer to a new job.

Other information systems will be important to ensure that you are paid, such as completing timesheets (records of the time spent on different activities), ensuring that your bank details are correct, and knowing your current leave entitlements, employment conditions, and that you can file claims for work-related expenses. Your supervisor should know all these systems and help you become familiar with them.

The first step on your journey from novice to expert professional engineer is to realise that the fastest way to find reliable information is to ask experienced people. These people will guide you past all the online material that can be irrelevant or untrustworthy, saving you lots of time. It is possible that your supervisor will know some sources, but even he or she will most likely refer you to other people in the company.

Therefore, you need to start building your professional network of people that can help you.

(13c)	Construct a personal database with systematically collected information on people familiar with local practices, materials, standards, suppliers, and service providers.
(13d)	Engage with engineers in other disciplines, service providers, suppliers, contractors, and experts to develop relationships in order to access local engineering knowledge. Demonstrate awareness of commercial and other interests influencing information provided.

Start your own professional contacts list in your phone, your work diary, and possibly in a spreadsheet table including:

> title, first name, family name, preferred name, mobile number (including country code starting with +), email address, Skype ID, and perhaps a personal or company website.

Include comments on the kinds of knowledge and expertise each person has.

Record the date and your interactions, e.g., "200901: showed me how to access Australian Standards collection." (I suggest writing the date in yymmdd format because you can then sort your spreadsheet table by date order.)

Make it a habit to record every person you meet in this database of contacts. It will be the most valuable piece of data you create in your career.

Initially, most contacts will be within the same firm. When you're new, most people you meet will be as helpful as they can. They remember when they started, and some will want to tell you all the things they wished they had known back then.

Maintain your relationships with these people. Make a point of meeting them from time to time, just for a short casual conversation, even when you don't actually need help.

This does not mean they will all become your friends. Workplace relationships are different from friendships, yet similar in many ways. They need to be maintained, and the easiest way to do that is to spend time with people, taking time to talk about non-work-related issues such as sports, families, vacations, movies, etc. or even to confess your growing awareness about how much you still have to learn.

This might seem awkward at first; you might be worried that the other person or their colleagues or boss will think you're wasting their time. Just having their contact details does not mean they will collaborate when you need their help.

Relationships lead to trust, and trust makes it easier to collaborate in the future. Someone you know well is much more likely to return a phone call when you need help or take the time to send a helpful and informative email reply.

However, you also need to extend your network beyond the firm.

For example, you will acquire much of your technical knowledge from specialist engineering product and software suppliers. These firms also provide application notes and design guides. Sales engineers have several motivations for educating you about the products and services they represent:

- You will not think of using a product you know nothing about—their employment depends on your company buying their products.
- To use most products, you need to learn how to apply them properly so that they perform as intended and provide you with reliable service.
- They know that you will recommend their products to others once you have had good experiences in learning how to use them and have seen them perform well in practice.

One of the fastest ways to expand your repertoire of technical knowledge is to attend specialist training courses provided by engineering product and service suppliers. Not only will you acquire useful and applicable knowledge, but you will also meet other novice and experienced engineers on these courses who will help you expand your professional network. The personal connections you can build on these courses tend to be far more valuable in the long term than the course material itself, which is often superseded by new technological developments within a few years.

Online courses can also be useful, but you miss out on the networking opportunities that come with face-to-face courses.

Next, you need to understand more about the different kinds of knowledge that engineers rely on. That comes in the next two chapters.

Chapter 10

Engineering knowledge

The key to engineering practice is understanding the nature of technical knowledge. Specialised knowledge, much of it technical, is the main attribute that distinguishes engineers from other people, and it also distinguishes the engineering disciplines from each other. For example, the technical knowledge of most electronic engineers is very different from the technical knowledge of most civil engineers.

Acquiring specialised knowledge is not that simple; most of it is not what was learnt at university. Much of it exists invisibly in the minds of engineers who are entirely unaware they possess it. Much of it, learnt informally on the job, can easily be taken for granted.

What, then, is this specialised knowledge, know-how, capability, or competence— this invisible 'currency' of engineering?

Knowledge and information

Philosophers have debated the concept of knowledge for thousands of years. Some focus on knowledge as a product of rational thought in the form of written statements of truth that are independent of any particular individual.

We are specifically interested in the knowledge held in the minds of engineers. I take the view that knowledge is 'justified true belief'. It is 'justified' in the sense that the person has taken personal responsibility to establish the truthfulness or validity of the belief.[1] The belief, therefore, has some basis in the experience of the person, and perhaps other people who, the person believes, are reliable informants.

Information, on the other hand, is data: the content of messages exchanged between people, machines, and systems. Information exists in many forms, such as documents, emails, text messages, CAD models, drawings, photographs, video and audio files, etc.

Over time, we use perception to construct knowledge in our memory. Remember from earlier chapters that perception relies on prior knowledge. It can be hard to transfer knowledge to other people because they need help to construct it in their own minds, and their different prior knowledge will result in knowledge unique to the individual. There is no such thing as an exact replication of knowledge; all learning is interpretation.

---

1 This discussion is adapted from an influential paper by Nonaka (1994, p. 15).

Information stored in computer systems can easily be replicated and transferred to other computers. It is hard to persuade other people to transform it into knowledge in their own minds.

Types of knowledge

Thinking about different kinds of knowledge helps to illustrate the different ways that knowledge can be constructed.

Explicit, codified, propositional knowledge

Most of the knowledge that you learnt in engineering school falls into the category known as explicit, or propositional, knowledge—also more broadly defined as 'codified knowledge'. A proposition is a simple statement that can be verified as being either true or false. These are examples of formal propositions:

> There are 242 pages in the geopolymer application field guide.
> Young's modulus, E, is the ratio of applied stress divided by the resulting strain in an elastic material.

Explicit knowledge is relatively easy to distribute. It can be written down in a language that appropriately educated people will understand, with a fair chance that their interpretation, based on their own prior knowledge, will align fairly closely with the intentions of the author. Explicit knowledge can be transmitted using symbols, such as words. In written form, as information, explicit knowledge can be transmitted without anything being lost. However, as soon as someone has to listen to or read it, and then interpret its words and symbols, some of the knowledge is inevitably lost or changed because of the variations in prior knowledge between individuals. All human language use is, in effect, translation between one person's ideas of what words mean and another person's ideas.

Explicit knowledge can be acquired by other people. It is not easy and is prone to errors and misunderstandings, as your own experiences at university can confirm.

Other kinds of knowledge are even more difficult to transfer.

Procedural knowledge

As you would appreciate from your studies, it is one thing to acquire explicit knowledge, but quite another thing to effectively use it. We can refer to the latter as procedural knowledge: knowledge that is needed in order to effectively make use of explicit knowledge, like a sequence of instructions. However, acquiring procedural knowledge requires one to practise the instructions until one no longer needs to refer to them.

You can study a textbook on mathematical statistics and acquire explicit knowledge that is relevant for the analysis of data resulting from experiments. Once you reach that stage of understanding, you can probably pass an exam.

However, you need experience in solving many practice problems in order to acquire the procedural knowledge that would enable you to competently apply statistical techniques to analyse your data so that you can confidently draw statistical conclusions from it.

Implicit knowledge

Implicit knowledge is knowledge that has not been made explicit (for example, by writing it down), but could be if needed. It includes things like knowing where the nearest bathroom is located. It might be written down on a plan of the building somewhere, but usually, we ask someone to show us or point the way. Most of us can remember where we need to go after the first visit, so we don't need to write it down.

The important thing to understand is that, unlike explicit knowledge, implicit knowledge can usually only be learnt with help from other people or personal experience.

Tacit knowledge

The term 'tacit knowledge' was devised by Michael Polanyi, who wrote, "We can know more than we can tell." An example is the knowledge you need to ride a bicycle or tie a bow with your shoelaces. This is knowledge that we acquire by practice—often frustrating practice—until one day, we get it right. From then on, it seems to come very naturally. Just for a moment, think of trying to describe how to ride a bike with words alone.

There are several different kinds of tacit knowledge. Riding a bike could be described as 'psycho-motor' knowledge or 'sensory-motor' knowledge. This is the knowledge that enables us to respond very quickly with movements that are appropriate for certain sensed conditions. Being able to walk without falling over is another good example of sensory-motor knowledge.

Tacit knowledge becomes embedded in our minds and bodies so deeply that we forget it even exists. We simply use it when we need it. It can only be learnt by practice and experience, often with the help of someone else. However, tacit knowledge is nearly impossible to verbally describe in a way that would be meaningful to someone trying to learn it.

Another aspect of tacit knowledge can be summarised as 'social knowledge', which is knowledge about how to behave and interact with other people in a given setting, according to local culture (Figure 10.1). The same person may interact in different ways in different settings—for example, at home, at work, with friends, with strangers, with authorities such as police, etc. It includes ways of talking and listening, what clothing to wear, body language, etc.

In engineering, tacit knowledge includes:

Ability to recognise objects, materials, defects, failure symptoms from appearance, sounds, smells, material feel, vibration, stains, and patterns of dirt accumulation;
Ability to recognise certain phenomena revealed by instruments such as oscilloscopes, thermographic imagery, and high-speed cameras—for example, electromagnetic interference that affect signals in an electronic circuit;
Visual understanding of drawings (e.g., circuit diagrams, P&ID diagrams) and the ability to visualise a three-dimensional artefact from a two-dimensional drawing or image. Also, the ability to know what is not shown in the drawing, such as

Figure 10.1 Unwritten tacit and implicit knowledge in engineering practice. Take a look at this construction site. How much of what you can see in the picture is shown in the drawings? Answer: not much, because the drawings show details of the finished building. They do not show all the steps needed to make it, such as in a LEGO instructions leaflet. For example, how did the builders know where to put the crane and how to construct the scaffolding? Those elements were not shown on the building drawings.

power supply connections that are customarily not shown on integrated circuit logic diagrams but are implied.

Appreciation of intrinsic beauty or the ability to create objects that will be visually attractive or beautiful to a chosen class of spectator or user.[2] Aesthetic knowledge can be based on other senses—for example, sound is the most important aesthetic sense in the realm of music recording.[3]

Ability to create a design that is not only visually attractive but also has appropriate proportions to make the best use of materials.[4]

---

2 Ewenstein and Whyte have discussed this in the context of building design (2007), and Ferguson has drawn attention to a similar appreciation for form in mechanical design (1992).
3 Susan Horning has described this in the context of the work of sound engineers (2004).
4 The modulus concept, in which a design for a particular application is obtained by adapting the proportions used in earlier designs, has been widely used since the early 19th century and can be traced back several thousand years in architecture (Guzzomi, Maraldi, & Molari, 2012).

Subconscious thoughts, imagination, and ideas that emerge without warning.

Tacit knowledge
   You don't know you need it.
   You don't know you know it.
   You don't know how to learn it.
   You don't know where to find it.
   You don't notice it even if you are looking at it.
   You don't notice that you have learnt it when you have.

Mathematics is often more tacit than explicit. Engineers seldom, if ever, apply the methods taught in university courses. However, we frequently use mathematical concepts when we discuss technical issues. We instinctively know, for example, that the highest point on a curve or surface has a horizontal tangent, a concept that originates in calculus. We also make instinctive mathematical decisions when choosing a simplified model that allows us to estimate design parameters faster.[5] All this comes from repeated practice exercises in engineering school that, at the time, seemed totally unrelated to practice. Many engineers think they never use their mathematical knowledge because they are unaware of their own tacit knowledge.

Embodied knowledge

There is another kind of knowledge built into the objects that make up our world. For example, the arrangement of a supermarket represents knowledge that has been developed over decades that makes life easier for shoppers, while also creating marketing opportunities for sellers. The arrangement of shelves, price tags, product packaging, and labelling, categorisations of products on the shelves, signs that direct you to 'cereals' or 'tea and coffee', the arrangement of checkout desks and cash registers . . . these are all manifestations of what we call embodied knowledge.[6] This is knowledge that is embodied in our world. Take roads, for example. Look at all the details of road design, line markings, reflectors, speed humps, kerb design, drainage, signs, and traffic lights; they all embody knowledge about helping people make effective use of the roads for safe transportation.

Embodied knowledge exists outside of individual people because it is embodied in objects. Sometimes it can be distributed by distributing the artefacts or images with descriptions.

Embodied knowledge is not necessarily easily accessible in the sense that a person can inspect an artefact and acquire all of its embodied knowledge. Many aspects of embodied knowledge may only be apparent to someone who has enough background knowledge. For example, much of the embodied knowledge in roads may not be

---

5 Gainsburg (2006) and Goold (2014).
6 Latour (2005, pp. 204–209).

apparent to a person with little or no experience of driving. Only an experienced road design engineer is able to recognise the more subtle aspects of a road's layout and hidden features like foundations and drains.

Contextual knowledge

Some knowledge is specific to a particular context, perhaps a particular organisation or even a specific workplace. Most people can recognise an electrical switch, even if they haven't seen that particular kind of switch before. However, knowing what a particular switch controls is a form of contextual knowledge. For example, knowing that a switch is on the wall of an American home and seeing that the movable part is in the uppermost position tells you that it is in the 'on' position. In many other countries, however, it would be in the 'off' position. This is an example of contextual knowledge: knowledge that is specific to a particular setting.

Knowledge transfer

Table 10.1 provides a summary of the different kinds of knowledge we need as engineers. They are not all mutually exclusive—for example, knowledge can be both explicit and contextual. What is important is being able to recognise how to acquire different kinds of knowledge.

Table 10.1 Types of knowledge

Knowledge type	External representation	How to acquire knowledge
Explicit	Logical propositions, written knowledge	Reading or listening, memorising, writing, and practice
Implicit		Experience with the help of others, reflection, with help of prior related knowledge and skills
Procedural[a]	Written or spoken instructions	Following instructions, practice.
Tacit—social, spatial, graphical, and visual language sensory-motor recognition		Imitation, practice, with the help of prior related knowledge and skills
Embodied	Artefacts	Examine artefacts, or detailed descriptions of them, with the help of prior related knowledge needed for interpretation
Contextual		Experience, conversation with experienced people, with the help of prior related knowledge needed for interpretation

[a]It is easy to confuse 'procedural knowledge' with a written procedure, which is an ordered list or a written statement of actions. The latter is merely a method of storing and transferring information. The former means knowing how to do something.

Acquiring new knowledge—learning

In constructing new knowledge in our minds, the learning process, we interpret perceptions of information in the light of our prior knowledge. Philosophers, education psychologists, and learning scientists have studied this process extensively.

One of the most important kinds of prior knowledge that we rely on is our language. As we shall see in the coming chapters, the idea of language as a convenient set of symbols with agreed-upon meanings is a bit too simplistic to explain human communication. However, it will do for the moment, provided we understand that we are continually learning about new meanings. We cannot take it for granted that the listener has the same understanding of a word or symbol as the speaker. Therefore, as we interpret new information in the light of existing knowledge, we have to understand that our prior knowledge base is continually evolving.

Knowledge develops in our own minds as a result of interpreting information that we receive every day. Much of our knowledge develops as a result of social interactions with other people, including parents, teachers, friends, and peers. As we discuss our own views and beliefs with other people, new ideas, and perspectives gradually emerge as a result of these interactions. This is particularly important within an organisation like an engineering enterprise, where the quality of the knowledge being developed and applied by people in the organisation is a critical factor in its overall success.

We will explore the social dimensions of engineering knowledge in the next chapter.

References and Further Reading

Ewenstein, B., & Whyte, J. (2007). Beyond words: Aesthetic knowledge and knowing in organizations. *Organization Studies*, 28(5), 689–708. doi:10.1177/0170840607078080

Ferguson, E. S. (1992). *Engineering and the Mind's Eye*. Cambridge, MA: MIT Press.

Gainsburg, J. (2006). The Mathematical modeling of structural engineers. *Mathematical Thinking and Learning*, 8(1), 3–36. doi:10.1207/s15327833mtl0801_2

Gainsburg, J., Rodriguez-Lluesma, C., & Bailey, D. E. (2010). A "knowledge profile" of an engineering occupation: Temporal patterns in the use of engineering knowledge. *Engineering Studies*, 2(3), 197–219. doi:10.1080/19378629.2010.519773

Goold, E., & Devitt, F. (2013). Mathematics in engineering practice: Tacit trumps tangible. In B. Williams, J. D. Figueiredo, & J. P. Trevelyan (Eds.), *Engineering Practice in a Global Context: Understanding the Technical and the Social* (pp. 245–279). Leiden, Netherlands: CRC/ Balkema.

Guzzomi, A. L., Maraldi, M., & Molari, P. G. (2012). A historical review of the modulus concept and its relevance to mechanical engineering design today. *Mechanism and Machine Theory*, 50(1), 1–14. doi:10.1016/j.mechmachtheory.2011.11.016

Horning, S. S. (2004). Engineering the performance: Recording engineers, tacit knowledge and the art of controlling sound. *Social Studies of Science*, 34(5), 703–731. doi:10.1177/0306312704047536

Latour, B. (2005). *Reassembling the Social: An Introduction to Actor Network Theory*. Oxford: Oxford University Press.

Nonaka, I. (1994). A dynamic theory of organizational knowledge creation. *Organization Science*, 5(1), 14–37. doi:10.1287/orsc.5.1.14

Polanyi, M. (Ed.) (1966). *The Tacit Dimension*. Garden City: Doubleday.

Trevelyan, J. P. (2014). *The Making of an Expert Engineer*. London: CRC Press/Balkema - Taylor & Francis, Chapter 5.

Chapter 11

Knowledge is a social network

Well, a lot of the problem solving here seems to be the people side. Getting who you need when you need and knowing who knows what . . .[1]

Many novice engineers think they have to know their technical 'stuff' or they could lose the respect of their peers and boss. Asking for help is a sign of giving up; a kind of cheating, a last resort. They believe that if they don't know it, one should look it up online, in texts, at the library, or even in Wikipedia (if they can get away with it), and they find it hard to admit that they don't know something.

A young mechanical engineer explained some of the things she found that she had to learn:

Practicality, how it all works, what a valve looks like, how you pull a pump apart. The next most important thing is where you fit in. Like others when I first started, I didn't think that I fitted in anywhere. I didn't have a job that no one else could do. As the scheduler, I had no idea how long it took to put in foundations. I had to ask somebody and figure out who to ask. I had a terrible lack of knowledge of actual 'stuff'.

Good engineers realise that knowledge is distributed in the minds of different people; it is not necessary to know everything; in fact, it is impossible. Knowing how to find someone to ask and how to get them to help is essential. Admitting ignorance is smart but takes courage. Every one of us has experienced feelings of utter ignorance and helplessness.

Most of the working knowledge needed by engineers lies in the minds of the people who work in the engineering enterprise. Vital knowledge lies with many non-engineers, including end-users. Accessing that knowledge effectively is the key to success in engineering.

By the end of this chapter, you will appreciate just how much more there is to learn. You will always be learning something new every day, which means that you will never stop learning. Engineering knowledge is limitless in its scope and detail, and there is no way of knowing what you are going to need around the next corner of your career.

---

1 (Korte, Sheppard, & Jordan, 2008).

Mapping knowledge

One of the first knowledge categories required in an engineering enterprise is an understanding of the different tools, systems, components, and materials that could be used—what they are called, what they look like, what each one does, its function in a particular context, and some understanding of its value or cost. This understanding does not include any detailed knowledge of how it is actually made. You do not need to know how a component or material is made in order to be able to use it. However, some knowledge of its manufacture and design can be helpful in understanding its limitations and identifying possible faults or failures. For example, it is not difficult to appreciate that a screw made from steel is likely to be much stronger than a screw made from plastic. However, it is more difficult to appreciate that a screw made from heat-treated alloy steel is likely to be much stronger than one made from mild steel without any special treatment, partly because they may look and feel almost identical.

Another related body of knowledge helps an engineer acquire the tools, machines, component parts, or materials—which firms supply them, in what quantities, and with what delivery times? Which firms hold them in stock locally? How much do they cost in a given quantity? How long do they have to be ordered in advance? Can the supplier deliver them or does transportation need to be arranged? Which firms provide helpful technical information, application knowledge, expert advice, or training? How is the machine, tool, component, or material safely transported and stored without affecting its performance when it is used? How long can the component or material be stored and in what conditions? How heavy is it? How can it be handled safely? Do these electronic components have to be protected from electrostatic charge? How long will chemicals that modify concrete properties retain their properties when stored in extremely high summer temperatures? Chapter 3 explains how a novice can start building this knowledge base while looking for work.

A third category of knowledge related to tools, machines, component parts, and materials is called 'detailed technical information', usually represented by technical data sheets and other technical information that enables an engineer to decide whether it will adequately perform the required function. This kind of knowledge also enables an engineer to predict the performance of an artefact or service that uses the component or material, how much it will cost, when it is likely to be available, the environmental and safety consequences of using the component or material, what kind of training will be needed by people who are going to work with the component or material, what kind of protective equipment they will need, and what kind of safety precautions must be taken. Apart from the knowledge of information specific to each item, engineers also need to know about combinative effects. Can tool X be used with material Y?

What happens if fluid F flows through pipes made from P, lined with R and S, controlled by valves made from V and W? A new fuel distribution system at a major Australian airport had to be scrapped and rebuilt after it was discovered that an anti-corrosion coating used in some of the pipelines could contaminate jet fuel. In cold weather, water had condensed from air trapped in fuel pipelines and tanks. The water became highly corrosive as it absorbed components of additives and contaminants in the fuel. Bacteria grew on the interface between the fuel and the water and produced highly corrosive waste products. Knowledge of components and materials must include knowledge of how to use them in combination with each other.

This takes us on to a fourth category of component and material knowledge: failure modes. Knowing how to anticipate and recognise failures is critical knowledge for an engineer to possess. An engineer must be able to predict the performance of a product or service before it is built or delivered. To do this, an engineer needs to ensure that the relevant tools, machines, materials, and components will be used well within their capacity. An engineer needs to anticipate the consequences of accidents and emergencies, even the consequences of failures in other parts of the product. Will a certain material have sufficient strength to withstand the forces applied during manufacture and assembly, as well as during transport to its final destination? Furthermore, when tools, machines, components, or materials fail unexpectedly, an engineer can use knowledge of failure modes to diagnose the causes and devise improvements to prevent future failures.

I have represented these four aspects of knowledge in Figure 11.1. Lines with arrows on the map indicate possible dependencies. For example, knowledge of logistics and procurement requires basic knowledge of the tools, machines, components, and the materials themselves, as does any detailed knowledge of properties, characteristics, and models. Knowledge of properties, characteristics, and models is essential in order to understand and predict failure modes. Sheaves of paper represent substantially written knowledge; the rest is constructed in the minds of engineers and skilled technicians, combined with their personal notes and work diaries.

I chose to represent categories of knowledge with clouds. A cloud is diffuse and has no clear edges or boundaries, just as knowledge classifications have fuzzy, ill-defined

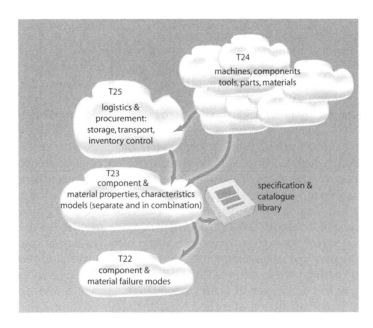

Figure 11.1 Mapping knowledge of tools, machines, components, and materials used in an engineering enterprise. Arrows convey typical associations linking fields of knowledge.

boundaries. It is difficult to capture a cloud, and, similarly, it is difficult to appreciate or fully summarise a particular area of knowledge. You can pass right through a cloud at night without even being aware of it. A person who lacks awareness can be surrounded by extremely knowledgeable people and not notice that they possess such valuable knowledge. Many Australian Aboriginal elders have a vocabulary of up to 300,000 words, ten times more than the average professor of English. Yet, for the first 200 years after they arrived in Australia, Europeans thought that Aboriginals were ignorant, uncivilised, primitive people. The fourth idea captured by the idea of a cloud is that you cannot tell how far into the cloud you are seeing from the outside, nor how thick the cloud might be. Unless you already have the required knowledge, it is difficult to understand how much you don't know. Fifth, even when you have some knowledge, when you are partly immersed in the cloud, you cannot see how much more knowledge you need to learn, just as you cannot see how much further into the cloud it is possible to penetrate.

Lastly, once you are fully in the cloud, it is very difficult to see the world outside. In the same way, once you become totally immersed in a particularly specialised aspect of knowledge, it can become more difficult to appreciate what is happening in the world around you. You can only see the outside world within the framework of the knowledge that preoccupies your mind.

Most engineering enterprises work with a range of products and services using different components and materials. Think about a layered map of clouds: each layer maps knowledge about one of the products, processes, or services.

So far, our map has included only four aspects of engineering knowledge: there are many more to come (Figure 11.2).

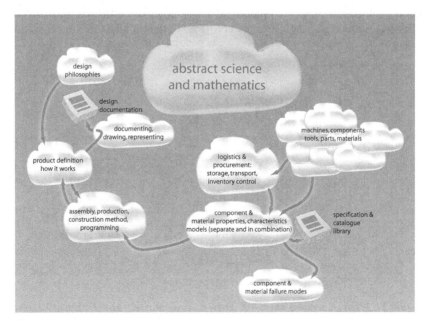

Figure 11.2 Knowledge map extended.

- Abstract knowledge, both mathematics- and science-based, from formal education
- Product definition, including how it works and how each of the components contributes to product performance
- Abstract model of an object, organism, structure, or physical system needed to predict performance
- Documentation: ways to represent the internal structure and operation of objects, structures, organisms, and physical systems
- Manufacturing and assembly methods
- Design for manufacture: ways to design products that result in economic manufacture and assembly in a given context
- Knowledge of components and materials
- Knowledge of component and material properties, both individually and in combination

The knowledge embodied in existing designs is also critical. For example, for a jet engine, here are some detailed aspects of embodied knowledge (Figure 11.3):

- Turbine blade shrouding, gas leakage, blade vibration, blade cooling, turbine inlet temperature, and pressure;
- Minimisation of turbine blade tip clearance, location of shaft bearings close to turbine discs to limit shaft deflection, and the use of turbine inlet nozzle vanes as the bearing support structure within the turbine housing; and
- Adjustment of seals and clearances to regulate pressure and flow of cooling air to the internal engine components, such as bearings and turbine discs, handling of engine axial loads, and the effects of cooling air pressure on lubricant flow in the bearings.

The accumulation of all this technical knowledge is a dominant constraint on the capacity of engine designers. It takes young designers many years to accumulate this

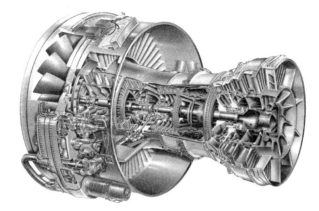

Figure 11.3 Large bypass ratio jet engine, typical on a contemporary passenger jet aircraft. The turbine at the rear of the engine is at the right-hand end of the picture (Rolls Royce).

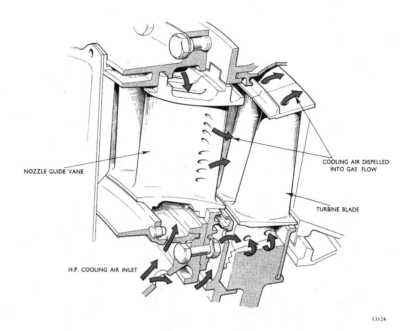

NOZZLE GUIDE VANE

COOLING AIR DISPELLED
INTO GAS FLOW

TURBINE BLADE

H.P. COOLING AIR INLET

13126

Figure 11.4 Nozzle guide vane and turbine blades showing cooling airflow. Modern turbine blades have many more holes and cooling air passages and operate in gas temperatures well above the melting point of the material (Rolls-Royce).

knowledge within the firm as they perform their design work; it cannot be obtained from other sources. Younger designers will acquire detailed knowledge in different technical aspects, but the design cannot be completed without taking the interactions between all these aspects into account (Figure 11.4).

This is not easy when the knowledge of all these different aspects is shared between many different specialist designers, and the design process is constrained by limited time.

Distributed knowledge

The research provides 37 types of specialist engineering knowledge and 27 types of general organisational knowledge. Two maps represent these, shown in Figures 11.5 and 11.6.

Note that these maps do not distinguish between the knowledge contributed by engineers and all the other people such as contractors, accountants, clients, end-users, and government regulators. As an engineer, it helps to know about all the different kinds of knowledge needed in an engineering enterprise, even though you will rely on other people for much of it.

The maps may seem complex. Then remember that the maps have to be replicated for every single product, service, and information package provided by an engineering enterprise. While there are areas of common knowledge between different products

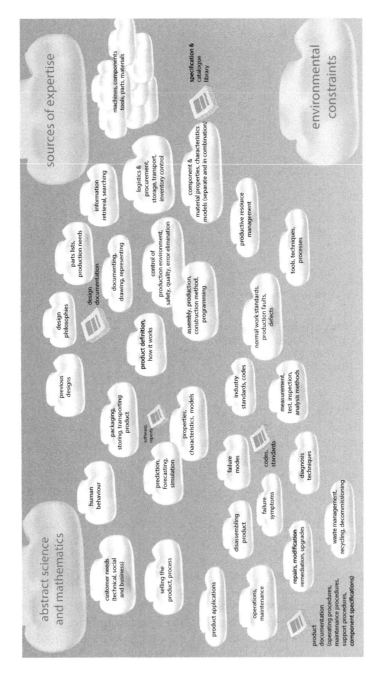

Figure 11.5 Aspects of technical knowledge used in engineering enterprises.

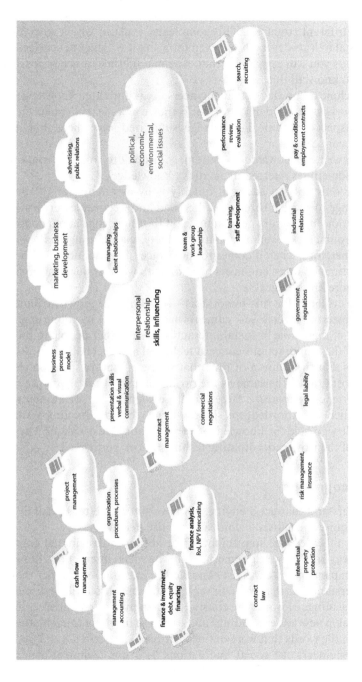

Figure 11.6 Aspects of organisational knowledge used in engineering enterprises. (RoI is return on investment; NPV is net present value.)

and services, there are also many differences. Think of these maps in layers, partly merged, partly separated, a different layer for every product and service. The layers must be subdivided yet again for producing the same product or service in different locations, or in different countries. This complexity explains why it is so difficult for any one individual to know everything, even for a single product or service.

By now, perhaps you can appreciate that even in a lifetime, it is not possible for an engineer to get to know all this 'stuff'.

Most of the technical knowledge that engineers use is unwritten knowledge learnt informally in the workplace. Written knowledge is shown as a sheaf of papers on the maps.

Nearly every aspect of technical knowledge is required somewhere along the way in nearly every enterprise.

This mostly unwritten knowledge is extremely difficult to transfer from one person to another. It is sometimes referred to as 'sticky knowledge'.

Therefore, the only way to access it quickly is by arranging for the people who have the knowledge to share it through skilled and knowledgeable performances. For example, production supervisors possess much of the knowledge about controlling the production environment, as well as managing safety, maintaining quality, and minimising errors. Therefore, when it comes to the choice of production methods and the design of production facilities, engineers need to work with production supervisors, as they are much more knowledgeable.

The strongest evidence for this hypothesis—that knowledge is distributed among people in an engineering enterprise—came from observations by young engineers that they spend most of their time in social interactions with other people in the same enterprise. Most of these interactions concern technical and specialised knowledge. The real surprise was that the amount of time that young engineers spent on social interactions was almost exactly the same as the proportion shown in many previous studies of more experienced engineers. Here was the evidence that social interactions are the means by which this knowledge is shared, largely by coordinating skilled performances by people who held the knowledge in their heads. This explains why engineers spend so much of their time coordinating technical work performed by other people.

Distributed cognition

By now, we understand that good engineers know that it's better to call in an expert than to try to acquire specialised technical expertise in a hurry.

Here, we take this idea one step further. The notion of distributed cognition implies that new ideas and knowledge emerge from the social interactions of people with different and complementary expertise, knowledge, and skills. Together they develop new shared understandings that could never have arisen without those discussions. As they do so, the meanings of words are constantly changing, often imperceptibly, as the participants develop new understandings and, through listening to each other, see new associations with the words that were out of their reach before.

While many engineers see social interactions as a non-technical aspect of their work, in reality, these social interactions involve highly technical discussions. Much of this occurs outside of working hours in relaxed social settings, and highly technical conversations can often be intermixed with jokes and speculation about upcoming sporting events and 'small talk'.

Figure 11.7 A graduate (female, centre) with abstract science, mathematics, and modelling knowledge has joined several other engineers in the enterprise that possess different aspects of technical knowledge. Between them all, if they collaborate, they can access all the knowledge they need.

Figure 11.7 shows part of the map of technical knowledge adapted from Figure 11.5. A female engineering graduate has just joined the enterprise. She's good at abstract science and mathematics, shown in the top left-hand corner. She can use mathematical models and computer software, but not in isolation. She needs help to come up with realistic estimates for the parameters that define her models.

There are several other engineers that cover the other aspects of the technical knowledge needed by the enterprise. Our young graduate will need to spend a lot of time working with them and figuring out realistic estimates for the mathematical models at the centre of her work. She soon learns that almost everything she needs to know is already known by someone else in the organisation; it is just a matter of finding out who that person is.

Engineering knowledge, therefore, is a social network.

References and Further Reading

Korte, R., Sheppard, S. D., & Jordan, W. (2008, June 22–26). A Qualitative Study of the Early Work Experiences of Recent Graduates in Engineering. *Paper presented at the American Society for Engineering Education*, Pittsburgh.

Trevelyan, J. P. (2014). *The Making of an Expert Engineer*. London: CRC Press/Balkema - Taylor & Francis, Chapter 5.

Chapter 12

Making things happen

In the last three chapters, I introduced the idea that engineering knowledge and resources are accessed through other people; that things happen in engineering through a social network. In this chapter, I explain how engineers make things happen when needed.

We learned about different kinds of knowledge in Chapter 10 and how acquiring knowledge is slow and susceptible to misunderstandings. Therefore, it is much faster and easier to arrange for people to collaborate and perform work that requires their specialised knowledge, skills, and resources rather than trying to learn for yourself. However, persuading busy people to do that, to make things happen when you need them is not easy.

The least effective way to ask a person to do something for you is by email or text message. There is a very good chance that nothing will happen. If something does happen, it's probably the wrong thing.

Most people in an engineering enterprise are working on many different tasks, more or less simultaneously. As a novice with perhaps two or three emails in your inbox each morning, you need to appreciate that people you need help from may have 50 or 60 issues vying for their attention, and hundreds, even thousands of unread emails. An email from you asking for information or help hardly registers at the bottom of their priority list and may even be forgotten in a few hours as it slips off the bottom of their inbox screen to join thousands of other unread emails.

Another way is to ask your supervisor to ask his or her supervisor to get the ultimate supervisor of the person who you need to do something to direct that person through their chain of supervisors. In other words, exercising management authority, working through an established organisation hierarchy that seems to exercise command and control throughout the organisation.

Think again.

First, it is very unlikely that the necessary messages are read, and perfectly understood by everyone up and down the hierarchy. Second, almost certainly you will be waiting a long time for anything to happen. Quite possibly, the wrong person will end up coming to do the work, and you will find that you need to ask all over again. In the vast majority of cases, you can persuade the person you need to collaborate with much faster on your own.

But how, you ask, can I, the most junior and inexperienced person in the organisation, possibly just walk up to someone and ask them to do something for me? Especially when they know far more than I do. I hardly know enough to ask them what needs to happen.

In answer to those questions, this chapter explains technical coordination, a special kind of informal leadership (Professional Engineering Capability Framework, Sections 2, 13):

Technical coordination, arranging for people to help you make things happen, preoccupies almost all engineers every day, from the first days of their careers, mostly without them ever thinking about it. It is the main method that engineers use to access technical information when they need it. Most often, when facing project deadlines, it is faster and easier to find someone with the knowledge, skills, experience, and resources to make a contribution quickly and skilfully. The right person may also have access to resources such as technical information, special tools, and equipment; through their networks, other people who you might not even know are needed to help.

Technical coordination seems to be something that all engineers do and takes around 30% of their work time. Even though some aspects resemble project management, at least in principle, they are distinctly different performances. Project management is a formal process that relies on documents and formalised relationships that we will discuss in another chapter. Technical coordination is an informal process that relies largely on undocumented social interactions and informal messaging.

I describe it as a four-step process, explained below, in which you gain the willing and conscientious collaboration of another person, which we refer to as a 'peer'. It does not depend on organisational authority or management hierarchy. In fact, conscientious collaboration is often more likely to occur in the absence of authority. People collaborate out of respect for one another.

Developing the capability for effective technical coordination is probably the single most effective way to build your engineering career, and emotional intelligence is an important attribute that helps. Employers value the ability to 'get things done' extremely highly (Figure 12.1).

Step 1: finding a peer

You will often know who is the best person to help you. It could be arranging for a specialist component supplier to provide samples for testing or evaluation. You could be coordinating work at a construction site, arranging to collect concrete samples for testing, and checking the installation of formwork for the next concrete pour. You may be coordinating the work of a maintenance contractor installing software upgrades at a telecommunications centre.

At other times, you may need to ask other people to suggest someone who has some specialised knowledge—for example, about different kinds of explosives. It helps to visualize a social network of technical expertise, as illustrated in Figure 12.2.

You, the novice, are the centre, and you now know a couple of people well, perhaps your supervisor and one colleague. Thick lines denote strong relationships, while thin ones represent weaker ones.

Start by reaching out to knowledgeable and experienced people with whom you already have strong relationships and ask them to refer you to others who have the expertise you need. If they were unavailable when you visited or phoned them, do not send text messages or emails unless there is no other way to reach them. If you do have to reach out by email, ask for another time to meet or call:

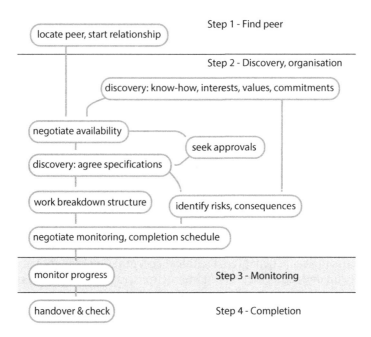

Step 1 - Find peer

locate peer, start relationship

Step 2 - Discovery, organisation

discovery: know-how, interests, values, commitments

negotiate availability

seek approvals

discovery: agree specifications

work breakdown structure

identify risks, consequences

negotiate monitoring, completion schedule

monitor progress   Step 3 - Monitoring

handover & check   Step 4 - Completion

Figure 12.1 Four-step technical coordination process: 'peer' refers to the person (or people) being coordinated.

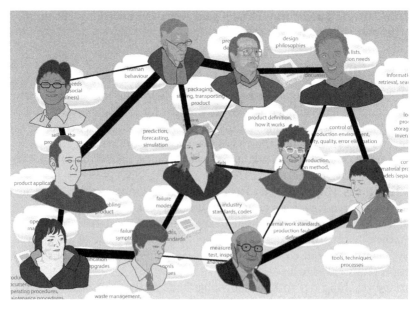

Figure 12.2 A social network of expertise. Thick lines represent strong, collaborative relationships.

I visited your office and tried to call you today, but you were not available. Please suggest a time that I could come and see you or call by phone.

They may suggest a phone call. That can work, but face-to-face encounters are much more effective, as explained in Chapter 7. If they can't help, ask them to suggest others who might be able to assist. Especially as a new person in the organisation, it is usually much better to meet people face to face for the first time.

It helps to continue building relationships with all the people you encounter in the search, as explained in Chapter 9. People with whom you have an existing relationship are far more likely to help you.

Step 2: discovery, organisation

The absence of authority requires that you agree with the peer on what they are expected to do and when. Even before this takes place, you should discuss current commitments with the peer and also consider the peer's know-how, values, and interests. This involves mutual discovery, preferably through a face-to-face discussion, so that you can confirm that the peer is the right person, and he or she will have enough time to complete the request on schedule.

It may be necessary to negotiate your peer's availability at certain times. He or she may need to renegotiate existing commitments and rearrange their schedule for the new task.

Unless the task is very short, it's a good idea for you or the peer to seek approval from, and at least inform, both your and the peer's supervisors, and also possibly the managers of any project teams to which they are currently contributing. There may be cost and budget implications. Apart from the cost of the time contributed by the peer, you will also need to spend time negotiating and monitoring the work as it is done.

Having cleared the way for the peer to devote time to the task, the next step is agreement on the details of the task between you and the peer. Again this is a mutual discovery performance because you will probably not know enough to be able to completely define the task without discussing it with the peer. The peer usually has special resources such as technical knowledge, skills, facilities, tools, or access to equipment—this is the reason why you need the peer to perform the task. You need to learn from the peer at the same time as the peer learns about the requirements from you. Together, you jointly 'firm up' or 'discover' the requirements as you improve your respective understanding of one another; gradually eliminating mutual ambiguity and negotiating the meaning of words you both use to define the task. It takes time to develop confidence from listening to the peer that the requirements are sufficiently well understood for them to start work.

If the task is complex, you may jointly develop a written work breakdown structure. In the case of a contractual arrangement, this kind of documentation is probably essential and may need external approval.

The last step in the organisation stage is to agree on a schedule, not only for the task completion but also for monitoring. You will monitor the task more or less frequently as the work progresses.

At the same time, at least privately, you need to anticipate issues that may arise during the task performance and assess risks that could affect the task performance schedule or quality. You should try to foresee unpredictable events and have some idea of how to handle these events if they occur, in conjunction with a series of backup plans. Of course, if there are health and safety issues, then it is essential to discuss these with the peer as well, and possibly others.

Step 3: monitoring—another discovery performance

Monitoring the task, often called following up, can be the most time-consuming step. It is a repetitive process, but it is also essential. Without monitoring, the peer may completely forget, or the work may be delayed because the peer focuses on other priorities. Another reason for monitoring is that you still carry responsibility for efficacy and safety. The peer may not have sufficient understanding to appreciate the potential consequences of the task, so monitoring and supervision are essential to ensure that intentions are enacted appropriately (Figure 12.3).

Monitoring starts with a prediction: the coordinator anticipates the current task status and progress, usually quite informally. Technical knowledge and prior experience help a coordinator make more accurate predictions, as they will know what to look for while following up.

The coordinator arranges to meet the peer face to face or at least to discuss the task by telephone. Monitoring by email is a last resort and is unlikely to be an effective way to rectify performance issues. The 'inspect progress' block in the diagram represents another discovery performance: a discussion with the peer about the task that results in the coordinator learning how the task is progressing. In doing so, both may also discover misunderstandings about the requirements and adjust their expectations accordingly.

Having assessed the actual state of the task, the coordinator then reflects on the difference between the predicted and actual state and, in doing so, adjusts expectations for future progress prediction.

The coordinator thinks ahead and anticipates consequences. For example, there may have been a misunderstanding about the requirement for quality; the peer may be taking much more time than was originally foreseen, ensuring that the work is performed

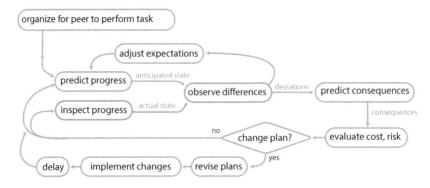

Figure 12.3 Monitoring process for technical coordination. Once the peer signals that the task has been completed, the final step—completion and handover—will begin.

to the highest standards through extensive checking. The coordinator may not have clarified the expected quality standard for the work. Also, engineers prefer to take time to seek the best possible results and check their work carefully. In the early phases of the project, particularly if the coordinator needs only a rough estimate from the peer, there may have been the expectation for the work to be done quickly.

The coordinator may have to think about different kinds of consequences and risks. If the task is being performed on a budget, he may have to renegotiate the budget or make allowances elsewhere. If the completion time is important, the coordinator may have to change the scope of the work or even find someone else to help move the process along.

If there has to be a change in plans, it takes time to discuss this with the peer, revise the current plans, and then implement changes. Bringing someone else in to help might sound like a good idea, but it almost invariably requires a significant amount of time; the two peers may also take more time to establish a productive working arrangement on the task. There is almost always some delay as a result of changing plans.

Mostly, this is all an informal process—monitoring hardly ever involves any documentation or notes.

The most important decision for the coordinator is to decide the frequency of monitoring. Too much monitoring can undermine trust—the essence of coordination is willing collaboration, and the coordinator has to display that confidence to the peer.

If the monitoring is too infrequent, however, then issues can arise that may seem unimportant to the peer but may be very important for the coordinator.

Another kind of issue is a technical difficulty that the peer feels confident in solving and continues to work through, relying on that confidence. The peer may try several different solutions before asking for help or even alerting the coordinator that there is a problem. The peer may need assistance from someone else, but that person may fail to respond as expected. The result is an unexpected delay.

As a rough guide, therefore, monitoring frequency depends on the likelihood of an unexpected issue occurring that the peer cannot solve by themselves, and the 'float time', which is the allowance in the schedule for unexpected delays. You should also take into account the possibility that if the peer cannot solve the problem, then it may be even more difficult for you to solve it; you may have to seek help from someone else.

Consider the supervision of a maintenance crew inspecting and replacing seals at bolted joints in a high-pressure gas pipeline. Technicians dismantle the joints by loosening and removing the bolts so that the internal seals can be inspected or replaced. Then, the technicians replace the bolts and tighten them using a torque wrench to ensure correct tension.

How do you ensure that the bolts are correctly tensioned? Or, as an electronics engineer, how can you tell that some delicate components have been appropriately handled in a clean room and protected from static discharge?

One option is to watch each technician, telling them not to proceed to the next stage until you have checked. This would be extremely time-consuming. It would be impractical to supervise more than two technicians working in the same location.

Instead, you would more likely allow the technicians to perform their job, knowing that they probably have far more experience performing this kind of work than you do.

Returning to the pipeline maintenance scenario, how can you tell that the bolt has been tensioned correctly? There is no difference in appearance. Perhaps you could use your own torque wrench to check the tightness of one or two of the bolts. However, you would not be able to tell whether the bolts had been overtightened, and then loosened, and then tightened to the correct setting. Would this matter? Yes, it would. Particularly in a high-pressure pipeline, ensuring that the bolts are correctly tensioned and not overtightened is essential to avoid fatigue damage to the bolts, as they are subjected to very high stresses in normal operation. Overtightening can permanently damage a bolt, causing invisible elongation, damage to the threads, and even a slight, but significant, enlargement of microscopic cracks in the metal.

Therefore, even though you are responsible for the repairs being performed correctly, you cannot actually observe every technician performing their work. Instead, you have to rely on the willing and conscientious collaboration of the technicians. You are relying on their willingness to perform the work correctly and conscientiously, taking care to ensure that their work is done correctly.

What we can see here is that the technical quality of the work performance affects the results in a way that may not be visible, even to someone watching at the time. As we have seen before, most engineering is performed under strict time and budget constraints: the work has to be completed according to an agreed-upon schedule and cost. Therefore, a great deal of technical work relies on both the conscientious performance of the work by individuals taking care to make sure that there are no mistakes and the levels of delegation and checking to minimise the likelihood of mistakes.

Trust makes a large difference here. Peers whom you trust require much less monitoring than people with whom you have not yet developed a trusting relationship. Trust translates into large time savings, reducing costs for everyone.

Contriving casual encounters

Naturally, most people are uncomfortable with excessive monitoring. Frequent monitoring can undermine trust and confidence. However, frequent monitoring is often the only way to keep a task on track, especially when the peer has numerous other distracting priorities to work on.

You will get to know that some people, even ones in whom you have the highest trust, will need frequent reminders, often the briefest phone call, just to keep the memory of your work sufficiently prominent in their minds.

Often, the monitoring only serves to keep you, the coordinator, in the mind of the peer, so they don't forget about the task they are performing for you. It is possible to do this in other ways, by contriving casual encounters.

An engineer described this technique:

> When I need to coordinate some extremely busy people, I try and position myself with a view of the corridor that leads to the bathroom or the coffee machine, preferably both. Then, when I spot someone who is likely to run late, I'll arrange to show up at the coffee machine or bathroom at the same time, seemingly by coincidence. Often, I will make sure I talk to someone else: all that is usually needed is for the other person to notice me and that makes them remember they're running late.

Step 4: completion and handover

The last phase in technical coordination involves careful checking by you and the peer to ensure that everything has been completed to their satisfaction. It may take time: you may both agree on testing with the expectation that any remaining performance issues identified as a result of testing will be rectified later (Figure 12.4).

It is a good idea to avoid a premature declaration that a given task is complete. Remembering that technical coordination relies on willing collaboration, it can be embarrassing to go back to a peer and confess that you did not check the work thoroughly enough when you initially accepted it. However, this can sometimes happen, even with the best intentions.

As with monitoring, your technical knowledge is essential to predict the consequences of accepting any remaining deviations, including the ultimate risks of doing so.

The most important aspect of the completion phase is to remember that task completion is not the only valuable outcome. As the coordinator, you and the peer will have strengthened your relationship. Relationships endure much longer than temporary coordination arrangements. Therefore, it is important to maintain that relationship. Otherwise, you will risk giving that person the impression that the relationship was only important in order to get a particular task completed.

This does not mean that you need to spend an excessive amount of time with the peer. However, it also does not mean that you can completely ignore the other person indefinitely or until the next time you require their collaboration. Take the time to visit the peer occasionally and view this as an important aspect of your work: building and maintaining trusting relationships.

Informal leadership, face to face

Technical coordination can also be seen as informal leadership. Most novice engineers, and many experienced engineers, reject the idea that they are exercising leadership.

Remember, as an engineer, you only achieve something by influencing the way that other people perform their work: manufacturing or assembling products, operating

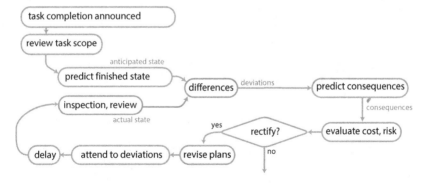

Figure 12.4 Completion and handover phase. The coordinator has to evaluate the consequences of any deviations from the agreed-upon task scope, as well as the cost and risks. If necessary, the coordinator will insist on deviations being rectified, accepting that there will be a delay and another inspection.

systems or processes, or delivering services to end-users. Influencing the way that others perform their work is a form of leadership. Influencing skills, therefore, are essential for engineers.

As we have seen in this chapter, successful influencing relies extensively on listening and perception skills. Go back to Chapters 4–8 regularly to evaluate your perception skills.

Face-to-face interaction with another person is nearly always the best way to start, but sometimes a carefully prepared piece of writing can convey ideas far better than a conversation where, perhaps, the other person is only half-listening and has several other pressing matters vying for attention. If so, use the conversation to alert the other person that you will send your ideas by email in writing, and then when they have read the email, you could meet to discuss them.

Social culture

Social culture can impose many complex constraints that can affect technical coordination. In many low-income countries, for example, it is well known that productive work requires the continuous presence of supervisors. Work quickly stops whenever the supervisor is absent, as explained in Chapter 17. Following up by telephone in South Asia, for example, can require phone calls several times a day.

In North American cultures, exercising influence without authority can be more transactional than in other countries. The peer may see the work as a kind of favour, to be repaid at some time in the future. In other cultures, the motivation comes more from mutual respect between the coordinator and the peer and the strength of their pre-existing relationship.

Practice exercise—knowledge network mapping

Map the knowledge network in your enterprise, at least within the circle of people that you routinely encounter. Remember that some of the people in your network, possibly many of them, work for other organisations, or maybe clients or end-users. Make a photocopy of Figures 11.5 and 11.6, and try to write names on each cloud to indicate people with the respective knowledge category. Draw lines of different thicknesses to indicate the strength of relationships. Repeat this after several months—you may be surprised to see how much it changes.

References and Further Reading

Blandin, B. (2012). The competence of an engineer and how it is built through an apprenticeship program: A tentative model. *International Journal of Engineering Education*, 28(1), 57–71. doi:0949-149X/91

Kendrick, T. (2006). *Results without Authority: Controlling a Project When the Team doesn't Report to You*. New York: AMACOM Books: American Management Association.

Rottmann, C., Sacks, R., & Reeve, D. (2015). Engineering leadership: Grounding leadership theory in engineers' professional identities. *Leadership*, 11(3), 351–373. doi:10.1177/1742715014543581

Trevelyan, J. P. (2014). *The Making of an Expert Engineer*. London: CRC Press/Balkema - Taylor & Francis, Chapter 9.

Chapter 13

Working safely

Many engineering activities involve health and safety hazards such as large, heavy objects, high-voltage electricity, extreme temperatures, radiation, concentrated energy, high speeds, and potentially harmful materials and fluids. However, even a seemingly benign office environment can pose significant hazards—for example, repeated strain injuries from using computers, psychological stress, and mental health problems, or in the event of a fire or earthquake.

This chapter provides some guidance, not only to keep you safe but also to help you eliminate or at least minimise health risks for others. As an engineer with specialised knowledge, you have a responsibility to anticipate health and safety risks and take action to protect others, whether or not it is your designated responsibility. You need to take this responsibility very seriously—to do otherwise can be a 'career-limiting move', especially if an accident occurs. However, it is more than just your legal responsibility. Failing to anticipate and prevent an accident that kills or maims people can leave you mentally tormented for years afterwards, thinking to yourself, "If only I had thought of that earlier!" Productivity may also improve if people in the organisation see that their health and security is taken seriously.

There is far more to safety than this brief chapter can describe. Therefore, the aim is to alert you to the main issues and direct you to more comprehensive information resources from which you can build your knowledge over time.

The best way to avoid safety hazards is always to eliminate them at the process design stage, and, in many cases, more time is spent considering safety issues during design than anything else. By learning about safety in engineering workplaces first, you will build up knowledge that you can apply to design work later in your career. Process safety has become a specialised career path for many engineers.

Identify hazards

Safety begins with an engineer's ability to anticipate the consequences of actions (or inaction). Foresight, the ability to predict many possible futures, is a vital attribute in all aspects of engineering (see the Professional Engineering Capability Framework, especially Sections 4a, 12, and 16). It takes time to develop comprehensive foresight—do not hesitate to ask others with more experience than you to check your predictions and spot omissions.

When thinking about safety, it is helpful to think of ways that an apparently safe environment could actually be hazardous, meaning that threats to health and safety are present or dangerous, and there is a significant likelihood of harm. A footpath beside a road with fast-moving traffic is a hazardous location. The road itself is an extremely dangerous location—an important distinction in thinking about safety.

Health and safety checklists are available online. Use them to create a hazards checklist for your workplace. You may need to ask others if any of the following are present:

- fast-moving objects;
- potentially harmful chemicals;
- pressurised gases, flammable dust, gases, liquids or solids, explosives;
- high-voltage or high-power electrical equipment (greater than standard mains voltage, or exposed connections with more than 50 Volts);
- harmful radiation;
- high-pressure liquid (greater than 20 bar); and
- noise greater than 80 dB (install a noise measurement app on your phone).

Obtain the materials safety data sheet (MSDS) for all chemicals or hazardous fluids in the workplace—for example, flammable gases. Read these and learn about containment and separation precautions that keep people from being exposed (Figure 13.1).

One of the greatest safety hazards is complacency, particularly among senior people in the organisation. This may not be difficult to assess. Ask when there was last a discussion on safety or a practice evacuation fire drill. Find out if harmful chemicals are properly labelled. Ask about advice on the health risks from using computers. Ask about the procedure to report a safety incident. Observe instances of corrosion or peeling paint that reveal a lack of attention to maintenance. Observe whether personal

Figure 13.1 Safety in a research laboratory. Safety was partly designed in at the start. For example, tubes carrying different gases are clearly marked and RCDs are in place to protect from electric shock. However, administrative controls are also needed by users—for example, maintaining accurate records of chemicals in use, ensuring that people who use the laboratory know how to handle the chemicals safely, and ensuring that PPE is worn.

protective equipment (PPE) is worn in hazardous areas, particularly by senior staff. If safety issues are not regularly raised, it is possible that senior people may not be paying attention to safety; others follow their example, and the resulting behaviour can lead to accidents or health problems. In an organisation with a strong safety culture, you can expect your attention to be on safety issues every day.

Identify hazardous events

A hazardous event is a foreseeable yet unpredictable event that is likely to have harmful consequences. For example, the entry of fast-moving flood water carrying floating debris can damage equipment, enabling poisonous chemicals to leak into the immediate environment. Another example: a truck driver who has lost his way drives unknowingly into a hazardous site because it is well-illuminated at night. He accidentally reverses the truck and damages valves on a high-pressure gas pipeline, releasing gas, which results in a fire or explosion. There are many possible hazardous events, and it is necessary to anticipate as many as possible. While all these events are foreseeable, their occurrence is unpredictable. In practice, most hazardous events result from human error and interpretation differences, while relatively few result from natural causes.

It is also helpful to ask lay people, not just technically knowledgeable engineers and technicians, about possible events and hazards. Cleaning staff, for example, will often notice things that others have missed.

Identify likelihood, consequences, and risks

When analysing safety hazards, it is necessary to identify the qualitative likelihood of each hazardous event. Select from about five possibilities, ranging from 'extremely unlikely' to 'frequently happens'.

In the same way, identify the consequences of each event, again using about five possibilities ranging from 'minor: possibility of injuries but none requiring medical attention' to 'catastrophic: many people killed or severe and permanent environmental damage'.

Then, each event is then given a 'risk rating' that combines the likelihood and consequences, again using about five possibilities from 'negligible risk' to 'catastrophic risk'. Engineers often use a risk matrix: a lookup table that assigns a risk rating to each combination of likelihood and consequences.

Risk control measures

Control measures are needed for all events, except for inconsequential risks. Events with the highest risk rating are considered first. A control, in this context, is some action that either reduces the likelihood of a hazardous event or reduces the consequences. For example, PPE usually reduces the consequences (personal injuries) following a hazardous event.

Engineers draw on their experience in assessing risks and devising controls, and often they will discuss the findings as a group. It might seem surprising that most of the time, only fuzzy qualitative descriptions are used rather than quantitative likelihood and numerical consequence scores.

This hierarchy shows the preferred control measures, with 1 being the first preference:

1 Eliminate hazard—as Trevor Kletz said, "Something you don't have can't leak";
2 Substitute hazard with something less dangerous;
3 Implement 'engineering controls' that reduce likelihood or consequences or both, e.g., shielding;
4 Administrative controls—procedures that reduce likelihood or consequences or both, e.g., rules, training, certification; and finally,
5 Personal protective equipment—eliminating or reducing harm to people.

Many control measures will be specified in regulations. Obtain the local occupational health and safety regulations. You should be able to demonstrate that you are familiar with relevant regulations for your industry and workplaces. If there are none, or if the local regulations are incomplete or only partly developed, consider downloading or studying regulations from an advanced country (e.g., UK guides and regulations at http://www.hse.gov.uk/, Australia guides and regulations at https://www.safeworkaustralia.gov.au/).

It is important to distinguish 'physical controls' (1, 2, 3 above) from 'administrative or procedural controls' (4, 5). An example of a physical control for a chemical is a secure container that will not break if dropped on the floor. Another might be arranging for all harmful chemicals to be contained inside sealed vessels and pipework. These are measures that keep people and harmful chemicals physically separated. Administrative controls include labelling the chemical containers, hazardous chemical signs on storage facilities and where the chemicals are used, procedures for handling chemicals, training for staff to follow these procedures, and record-keeping. Administrative controls are also measures that keep people and harmful chemicals separated, but rely on people to observe signs, follow procedures, and obey rules.

In practice, especially with very harmful chemicals and other extreme hazards such as high-voltage connections, physical and administrative controls are both used in combination. For example, in situations where floors can be wet, electrical equipment should either be eliminated or protected with residual current devices (RCDs). While an RCD is a physical control, it might need to be connected to the electrical power outlet by a user. Administrative controls, therefore, could include regular random checks by supervisors to ensure that RCDs are being used properly and informative notices explaining how to use them in appropriate languages.

Using the risk matrix approach outlined above has sometimes encouraged engineers to consider extremely unlikely events as insignificant risks and therefore defer implementing control measures. Less serious safety risks that are more likely to occur were given higher priority. Now, in Australia and other advanced countries, control measures are needed for every event with major consequences, no matter how unlikely they seem to be. The reason for this change is that time and again engineers have disregarded the possibility of extremely unlikely events actually happening. Most people, not just engineers, underestimate the probability of low-frequency (unlikely) events. Tightened rules represent lessons learned from major incidents, such as the Piper Alpha oil platform fire in 1988.

Keeping detailed records is an essential administrative control. History shows that major accidents are almost always preceded by minor incidents or near misses. Making it safe and easy to report incidents and near misses can provide data that suggests more effective control measures that can prevent a major catastrophe.

First steps

Learn to anticipate the foreseeable but unexpected.

Learn about safely using computers and about mental health in the workplace.

Enrol in a first aid course and, unless training has already been provided, learn how to safely lift heavy objects. You never know when you may have to help lift some heavy equipment.

Take advantage of opportunities to attend courses on electrical, chemical, and other relevant safety practices. If your enterprise demands that people work in extreme conditions—heat, cold, at heights, underground, even shift work—learn how these environmental conditions affect health and behaviour.

Cultural influences

Safety culture, habitual ways of thinking, and acting on health and safety hazards, has a pronounced effect on practice. For example, it is not uncommon to hear influential oil and gas business leaders explain that "there has to be a balance between safety and production." This is used to justify limiting spending on safety precautions. Companies with leaders espousing similar attitudes have tended to experience major disasters: BP is one such company. They seem to forget something that is particularly true in that industry: production is only possible with safety. Disasters, when they do occur, usually stop production completely and often bring very large claims for compensation. The reputation of even a large and well-known company can be lost or badly damaged. BP was close to bankruptcy after the Gulf of Mexico Macondo well blowout disaster in 2010. This was because many suppliers started demanding payment in advance before delivering goods and services; they were scared that BP might go bankrupt and, as a result, BP almost ran out of cash.

Many countries have cultural attitudes that can conflict with a strong safety culture. For example, in some Muslim countries, many people subscribe to a belief that Allah (God) will determine their last day; they believe that no person can change that, so safety precautions are of limited value. In many developing countries, there is a perception that 'labour is cheap' and easily replaceable, so spending money on safety precautions is not needed beyond the minimum regulatory requirements, which can seem lax by advanced country standards.

As a young engineer in these situations, it can be challenging to promote safety in the prevailing culture. However, there are very good reasons for doing so, quite apart from one's conscience and moral duty to care for all people.

The first reason is your personal safety: if the organisation cannot maintain a reasonable safety culture, there is a greater risk to your personal health and safety.

The second reason is economic. Even in low-income countries, labour is not cheap (explained in Chapter 17). Production usually depends on competent, highly paid supervisors who must be continuously present and visible. In typical engineering enterprises, the largest costs are materials, energy, and fixed costs, such as machinery and land. Losing an experienced worker due to an accident can significantly reduce utilisation and cause material wastage, adding considerable extra costs.

The third reason is that culture can nurture misunderstandings. Like all other major religious texts, the Qur'an, accepted by Muslims as the literal words of God, instructs believers to receive wisdom and also to respect the sanctity of life. Believers, therefore, have an obligation to learn how to avoid harming themselves and others.

The fourth reason is reputation. Most international supply chains are now audited for ethical and sustainable practices. A good safety record can make a crucial difference in winning orders from major customers, and also for securing finance for working capital to handle large production contracts. It can also reduce the cost of insurance.

Human behaviour

No matter how well physical control measures have been designed and built, safety relies on human behaviour above all else. Even the best physical controls must be disabled from time to time for maintenance or modifications.

Understand that people do not necessarily follow the rules created for their own safety without visible and continuous enforcement. Signs warning of hazards, particularly when seen repeatedly day after day, eventually fade from consciousness. Workers will violate strict safety procedures. As time passes after a major incident or safety drill, complacency will inevitably arise.

As an engineer, you must anticipate that people will sometimes engage in risky behaviour or will sometimes be intoxicated by alcohol or other drugs.

It will be your job, more than most other people, to do your best to keep people safe from harm, despite their behaviour. As an engineer, you may be designing administrative controls: rules and procedures for people to follow.

Careful observation of human behaviour can be one of the best ways to maintain safety. Workers who consistently decline to use PPE, such as safety goggles, may indeed be smart. Poorly designed PPE can interfere with vision and hearing or increase discomfort and fatigue, raising the chances of accidents. Procedures that are too complex or onerous will be bypassed. If you notice this behaviour, take the time to gain the confidence of workers and listen carefully to their explanations before making changes. Supervisors and shift leaders can be the most helpful, as they are much more likely to be able to influence the behaviour of workers than you can.

Ensuring reasonable safety may demand a large proportion of your time and attention. Furthermore, you may find opposition from within your own company. Attention to detail, knowing regulations, understanding business economics, appreciating reputation factors, and persistence will often pay off in the end.

Sometimes, however, the risk to your own health and well-being may tip the balance in favour of looking for alternative employment opportunities. In the meantime, while you're looking, tackling difficult safety issues will provide invaluable experience for your future career.

References and Further Reading

Bea, R., & Deep Water Horizon Study Group. (2011). Final Report on the Investigation of the Macondo Well Blowout. Retrieved from: https://ccrm.berkeley.edu/pdfs_papers/bea_pdfs/DHSGFinalReport-March2011-tag.pdf.

Chernobyl disaster: https://en.wikipedia.org/wiki/Chernobyl_disaster.

Kletz, T. (1991). *An Engineer's View of Human Error* (2nd ed.). London: Institution of Chemical Engineers, VCH Publishers.

Levy, M., & Salvadori, M. (1992). *Why Buildings Fall Down.* New York: W. W. Norton.

Petroski, H. (1985). *To Engineer Is Human: The Role of Failure in Successful Design.* New York: St Martin's Press.

Petroski, H. (1994). *Design Paradigms: Case Histories of Error and Judgment in Engineering.* New York: Cambridge University Press.

Reason, J., & Hobbs, A. (2003). *Managing Maintenance Error.* London: Ashgate.

Standards Australia. (1999). *Australian and New Zealand Standard 4360:1999 Risk Management.* Retrieved from saiglobal.com March 2013.

Waring, A., & Glendon, I. A. (2000). *Managing Risk: Critical Issues for Survival and Success in the 21st Century.* London: International Thompson Business Press.

Warren Centre for Advanced Engineering. (2009). Professional Performance, Innovation and Risk in Australian Engineering Practice (PPIR). Retrieved from Sydney: https://thewarren-centre.org.au/project/professional-performance-innovation-and-risk/;

Chapter 14

Making big things happen

Reputations depend on delivering on promises. Once you can do that, people will trust you to spend their money. This means that you must be able to manage projects.

Engineers organise and manage projects to make big things happen.

Most technical work is largely autonomous and relies extensively on social interactions. It is also diverse. You may have 60 or more simultaneous issues to deal with: requests for assistance, coordination of other people, decisions that must be made, collecting information to help with decisions, email correspondence, and administrative issues. What we do each day, the order in which we do it, when we choose to start, what we remember to do, and what we forget, when things get interrupted by others, when we forget to return to interrupted tasks—all these contribute to uncertainties. There are also mistakes: unconscious slips made without any awareness, deliberate incorrect actions taken because of incorrect perception, deliberate incorrect actions taken because of interpretation differences, deliberate incorrect actions taken because of habit, failure to acknowledge incorrect actions and take corrective action in time, and incorrect actions taken with the intent to deceive and keep them hidden. Delivery, therefore, depends on the interpretations and subsequent actions of many people, working with incomplete information and lots of uncertainty.

Fifty-five per cent was good enough to pass at university, but as an engineer, even 99% may not be good enough. How is that possible with so much more uncertainty, especially as there is never enough time to read every email in your inbox?

You can learn. Read on.

Project management is a family of methods that engineers use to limit the consequences of all this uncertainty. It includes

- explaining objectives
- skill development and training
- organisational procedures and checklists
- risk management
- project planning
- managing interruptions
- technical standards
- monitoring
- appropriate tools
- appropriate organisational culture
- quality assurance

All these depend on an elaborate system to manage documents: collective memory for decisions and action plans.

'Project Management' emerged in the 1970s when people applied engineering methods to organise projects in many other contexts. From new product launches organised by marketing and sales teams to organisational restructuring by management consultants, these systematic engineering methods proved to be remarkably successful. The US-based Project Management Institute now provides education and qualifications worldwide based on these techniques.

Unfortunately, engineering project delivery performance has declined since then, according to data gathered by the leading international firm that monitors the success of large engineering projects. Data on more than 15,000 projects have revealed that less than one in three projects over USD 1 billion managed to provide greater than 50% of forecast investment returns from when the projects were approved. Around one in six of these projects were complete failures, abandoned by their owners. While smaller projects succeeded more often, the overall performance can only be described as appalling. This is why engineers have such tarnished reputations among company owners, senior managers, and governments.

Many engineers are surprised when they hear this: they work extraordinarily hard to deliver results on time and budget for clients. However, they seldom see projects from the owner's perspective. While many parts of a large project might have progressed according to plans, the overall project can still fail. The owners often remain silent, preferring to draw attention to successes instead. Perhaps because owners are reluctant to allow open investigations, we still only have very limited understanding about these failures and little evidence of improvement.

Engineering practice research has revealed some relatively neglected aspects of engineering project management, as presented in the next two chapters:

a   difficulties converting information into knowledge
b   technical specifications
c   value perceptions

A deeper understanding of these aspects might help engineers improve project delivery performances. I sincerely hope it does, and also that you can help to demonstrate that in your career.

Information, knowledge, and diversity

Project organisation relies on extensive information recorded in documents and information systems. As explained in Chapters 4, 6, 10, and 11, information can readily be moved from one computer to another. However, project organisation depends on people acquiring knowledge from documents and information systems so they can make appropriate decisions when needed. It is seldom recognised in project management texts that converting information into knowledge in the minds of people is difficult and presents special challenges.

Another challenge is that it is difficult to ensure that all the information is consistent and as free from errors as humanly possible.

One of the most effective ways to manage uncertainty is systematic checking, exploiting diversity in perception, interpretation, thinking, and action. Different people see the same things differently; that is why it's best for documents to be checked by someone other than the authors. The more the checker thinks differently from the author, the more likely the checker will be able to notice mistakes in a document. The mere expectation that work will be systematically checked by senior engineers can improve performance just by raising expectations on the expected standard of work. Encouraging diversity within an engineering enterprise is likely to expose mistakes and misunderstandings earlier, resulting in fewer costly consequences.

Document-reading skills are especially valuable in this context, as explained in Chapter 6.

Project life cycle

Every project starts with discerning needs and negotiating possibilities, along with securing funding and regulatory approval. In practice, funding and regulatory approval precede each of the main stages through a 'stage gate' decision-making process that is usually unique to a particular enterprise (Figure 14.1). For more, see *The Making of an Expert Engineer* (Chapters 10 and 11).

Design and technical problem-solving come next, to the extent that they are needed. In practice, design and problem-solving are often interwoven with analysis and prediction processes in a series of iterative refinements. Even the requirements may be renegotiated in the early stages.

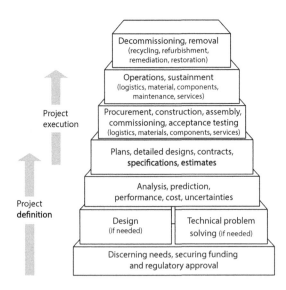

Figure 14.1 Sequential activities or stages in an engineering project, starting from the project definition phase at the bottom, explained in the text below. Not all activities have been shown. Different engineering ventures naturally place different emphasis on each stage. For an engineering consultancy, the 'product' is usually information, often construction drawings and specification documents.

Detailed planning and preparation of all project documentation is usually the last step before the critical final investment decision (FID) is made to proceed with the rest of the project. Typically, by the end of this stage, between 5% and 10% of the project budget will have already been spent. From this point, everything gets much more expensive.

Procurement and logistics, manufacturing, installation, commissioning, and acceptance testing use up the bulk of the capital investment in any project. Predictability is the key in this stage: engineers are under a great deal of pressure to make sure that everything is completed safely and on time, within the planned budget, and delivered with the required technical performance.

Only in the last stage does all this investment start to provide some useful value for people through the tangible products or services provided. This is when payments start to flow back to the investors who provided the funding for the venture. The product must work reliably for its expected service life, and often much longer. Maintenance plays an important part in achieving this, which means keeping everything working so that the predicted performance is achieved throughout the service life of the product or process. Ultimately, there will be a decommissioning step when the artefacts are removed for reuse or recycling. In the case of a manufactured consumer product, this can happen throughout the life of the product.

Figure 14.2 shows the project activities enclosed within a coordination ring that continually guides the implementation steps towards the intended objectives. A web of social relationships coordinates each of the technical activities. The ring consists of informal (and often invisible) processes grouped on the left and their formal equivalents on the right. In a way, the formal coordination processes are also invisible: they are often regarded as 'non-engineering' activities.

Notice how design and technical problem-solving are relatively insignificant aspects in these diagrams. Technical problem-solving is avoided as much as possible; it is usually much more preferable to use solutions that have been tried and tested in the past rather than devising new ones with uncertain efficacy. In many projects, engineers reuse designs from previous projects as frequently as possible. Once again, this saves time and reduces uncertainty.

Notice how the coordination ring also acts as the foundation for the whole endeavour.

The coordination ring involves continual interaction between all the participants, including the client(s), financiers, engineers, contractors, suppliers, production and service delivery workers, technicians, regulators, government agencies, the local community, and special interest groups.

In the coordination ring base, work starts with negotiations on constraints, even before funds have been committed. Constraints include

- capabilities of suppliers, production capacity
- technical requirements
- schedule
- regulatory requirements
- health & safety requirements
- environmental impact, emissions
- reliability requirement, client's maintenance capacity
- client's financial capacity

- external financier(s) requirements
- tolerance for uncertainty
- intellectual property

These negotiations shape decisions on funding at each stage of the project.

On the formal side of the coordination ring, we find engineering management systems, including project management, configuration management, environmental management (e.g., ISO 14000 series), health and safety management (e.g., ISO 18000 series), quality management (e.g., ISO 9000 series), asset management (or sustainment, e.g., ISO 55000 series, formerly PAS-55), document management, and change management. You will encounter these systems throughout your career.

On the informal side, there is a continuous renegotiation of meaning as the project team comes to understand more about the requirements and constraints. Different participants initially attach their own meanings to the terms used to describe every aspect of the project, but as the project proceeds, these differences must be resolved or at least understood and acknowledged.

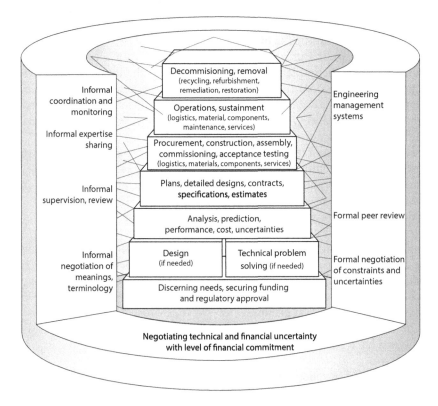

Figure 14.2 Project management and coordination. The stack of project steps is surrounded and supported by a coordination ring that guides the project and manages all the uncertainties introduced by human performance. Coordination links with the world outside of a given project also occupy much of an engineer's effort; these links are not shown on the diagram but are equally significant.

For example, there may be differences in the way that specifications are interpreted. Many people think a specification is a non-negotiable statement of requirements: components cannot be accepted unless they pass all tests at the required level of performance. However, others may think this only applies to production items. Pre-production versions of certain components can have lower performance. Some people may understand a specification to be 'elastic,' meaning that as long as the essential requirements are met, other non-compliances could be negotiated away in the form of a price discount, if and when they became apparent.

Some engineers talk about 'reliability' issues that others consider manufacturing 'quality' problems. Different individuals involved in the coordination ring construct their own knowledge and understanding in different ways, which can make the process of sharing that knowledge a lengthy and difficult one at times.

The most important idea in this discussion around Figure 14.2 is that a major component of engineering activity needed for a project is project management and coordination work. Many engineers spend nearly all their time on this; often, the time spent on calculations, design, and problem-solving can be tiny, as little as 2%. Many engineers think that their project management and coordination work is 'not real engineering' because it's not taught in engineering schools and seems to be non-technical. However, these aspects of engineering work strongly influence project delivery performances, which must improve if we are to restore today's tarnished image of engineers.

Project planning

Managing a project can be described as a three-step process:

1 planning and organisation,
2 monitoring progress, and
3 completing the project.

Every task within a project also requires these steps.

A well-managed project needs a detailed plan, the first of many project documents.

Even the best project managers don't get everything right the first time. Instead, nearly every document you create is a living document that evolves and is progressively elaborated on during the course of the project. At each stage, you, your colleagues, and often the client and other stakeholders will get a clearer idea of the requirements. Some documents are still being elaborated even at the very end of a project.

Always look for ways to simplify the requirements without losing detail, and keep documents as short as possible (Figure 14.3).

Negotiate and define the scope of work, calculate the time schedule

This is the single most important job for a project manager: negotiating and reaching an agreement on defining the scope with the client and other stakeholders (see below). Some projects do not involve a paying client; they may be sponsored by someone with

Figure 14.3 Project definition and planning phase. While there is a clear sequence, in reality, the process is at least partly iterative. As significant technical issues and uncertainties become clearer, and as more investigation by the project team provides insight, it may be necessary to renegotiate the scope and risk-sharing. It may even be necessary to revisit some of the planning once the project gets underway, as shown in Figure 14.6.

the appropriate authority in your own firm. The sponsor instigates the project and can, therefore, be treated as equivalent to the client.

Mostly, the sponsor or client has only a sketchy idea of the requirements. In the early phase of the project, therefore, your main role will be to clarify requirements and specify the necessary details.

Project scope is defined in one or more documents that provide a succinct definition of everything that is to be accomplished by a project.

While everything else can be progressively elaborated, be extremely cautious about elaborating anything in the project scope. This is called 'scope creep,' a series of small extensions to the scope, each one being apparently insignificant. Adding anything to the project scope almost invariably means increasing the cost and time to complete the project. Finalising the project scope definition right from the start is one of the most crucial skills of a good project manager.

The project scope enables engineers to prepare a work breakdown structure (WBS), a complete list of work activities to be performed by everyone concerned with the project. Each activity should normally take at least one day to complete and not more than 20 days (four working weeks). Otherwise, it can be difficult to monitor progress.

Another important document, the Bill of Materials (BoM), lists every item needed for the project and how it is to be obtained, transported, and stored.

By specifying all the constraints that determine, for each activity, what must be completed in advance, a computer can calculate the expected time schedule and display this as a Gantt chart (Figure 14.4).

This aspect of project management is almost routine and can be easily arranged with one of the many software packages that are now available.

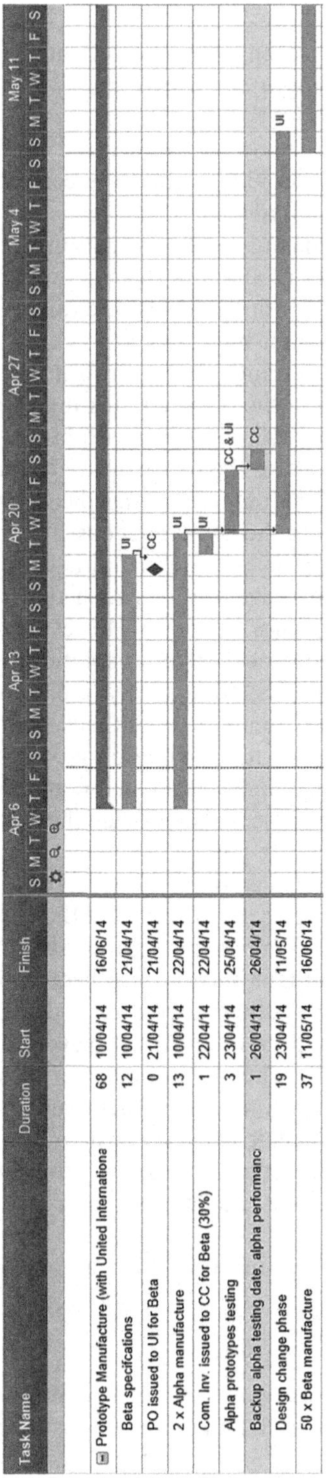

Task Name	Duration	Start	Finish
⊟ Prototype Manufacture (with United Internationa	68	10/04/14	16/06/14
Beta specifications	12	10/04/14	21/04/14
PO issued to UI for Beta	0	21/04/14	21/04/14
2 x Alpha manufacture	13	10/04/14	22/04/14
Com. Inv. issued to CC for Beta (30%)	1	22/04/14	22/04/14
Alpha prototypes testing	3	23/04/14	25/04/14
Backup alpha testing date, alpha performanc	1	26/04/14	26/04/14
Design change phase	19	23/04/14	11/05/14
50 x Beta manufacture	37	11/05/14	16/06/14

Figure 14.4 Small part of a project Gantt chart.

Specifications

Technical specifications are no longer mentioned in project management courses. However, they are critical for engineering projects.

Why?

Technical specifications define how the quality and completeness of technical work will be assessed. Engineering activities cannot be certified as completed unless the outcomes comply with the required elements of relevant technical specifications and standards. Therefore, project management and costs in an engineering context entirely depend on technical specifications and how they are interpreted.

For example, the way the operating temperature range is defined for equipment influences cost. Typically, this might be stated as $-55°C$ to $125°C$, a common military standard operating temperature range. However, if it is stated as $-42.5°C$ to $52.5°C$, the equipment cost might be ten times higher. Testing equipment that complies with this latter specification might require environmental test chambers with $0.1°C$ precision temperature control because the temperature is specified to $0.1°C$! These test facilities, required by both the manufacturer and the organisation that certifies compliance, may add considerably to the equipment cost.

What the specification rarely, if ever, describes is the end-user's needs. Engineers interpret these needs and write specifications such that the final solution will meet those needs, in other words, 'fit for purpose.' However, the engineers designing the solution may conceive of far more cost-effective solutions if they can fully understand the ultimate needs or requirements. Almost invariably, that means finding potential end-users and listening carefully to understand their needs.

The specifications will also refer to drawings or, more likely, digital representations of the object to be created—encompassing most, if not all, aspects of the design. These representations of the complete object also provide the details needed to construct the bill of materials and WBS.

On many fast-moving projects, it is common to include a 'contingency' in the budget: an amount of money to cover items not specified at the start.

So, the next logical question is: how is a specification structured as a written document?

For this, we need to understand the purpose of a specification: it defines what is acceptable and how we can determine whether the particular deliverable is acceptable or not.

Although the client does not need to understand every detail contained in the specification documents, the people who are responsible for delivering the artefact, material, service, or information will need to understand. As an engineer, one of your primary responsibilities is to decide the level of detail required for the specification document. One way to reduce detail in a specification document is by referring to national or industry standards where details of requirements and testing methods are already available.

There are two basic types of specifications: test and method.

Test specification

The first type of specification is known as test specification: we describe the testing and inspection that will confirm that an objective has been achieved (in other words, the result complies with the specification) or has not been achieved (because the result does not comply with the specification).

For example, we might be required to construct a beam. The specified acceptance test might consist of a defined loading on the beam and measurement of its deflection at a given point. Hence, the specification might define

a   how the beam is to be positioned and supported,
b   the applied load(s) and how the loads are to be applied (e.g., a testing machine, weights, attached steel ropes, etc.),
c   allowable deflection limits in certain directions and rotational movements at given positions on the beam,
d   the length of time for which the load is applied, *
e   maximum and minimum temperature and relative humidity limits for testing, *
f   storage conditions for the beam prior to the test, * and
g   the maximum allowable wind (if the test is done outside).

Items marked with an asterisk (*) are important for non-metallic structures, particularly fibre-reinforced composites, plastic materials, or natural materials such as wood.

If possible, it is preferable to use standard tests so the results can be compared with previous tests. For example, the American Society for Testing and Materials (ASTM) is an international standards organisation that develops and publishes voluntary consensus technical standards that focus on testing methods.

The weakness of a test specification is that the testing is only performed once, when the artefact, information or service is provided. However, we might need an artefact that performs reliably for 30 years with only annual inspections and minor maintenance, such as repainting. How can we be confident that an artefact that passes an acceptance test on delivery will still be fit for purpose 20 or 30 years later?

One option is to incorporate what is known as accelerated ageing into the test specification. An artefact intended for outdoor application in a coastal environment could be subjected to alternating high and low temperatures, high and low humidity, intense ultraviolet radiation, salt spray, and vibration, perhaps simultaneously. The degradation that might take a decade or more in the actual location might be reproduced with 2 months of testing by using accelerated ageing.

This kind of testing, however, has limitations. Another kind of specification, method specification, provides an alternative approach that can help overcome this weakness.

Method specification

The second kind of specification is known as method specification: we describe the method, process, or procedure by which an objective is to be achieved. A method specification will often include a detailed specification of the tools, materials, and monitoring of the production process.

A method specification gives us confidence that an artefact will perform as intended by ensuring that it is constructed in a known manner by suitably trained and experienced people, it is made from materials of known quality, and it is inspected during the manufacturing processes using standardised techniques.

Sometimes, a specification requires a combination of these two approaches. For example, a method specification may require certain intermediate acceptance tests to be performed during construction, manufacturing, or assembly processes.

Inspection and testing plans

Apart from the specification documents, the project will also need inspection and test plans (ITPs). ITPs require careful thinking and writing. Along with specifying methods and tests, an ITP must also define what is to happen if any of the inspections or tests reveal non-compliance.

Here is an example of where that was not done. The hydraulic control unit shown in Figure 14.5 was subjected to a long sequence of acceptance tests before leaving

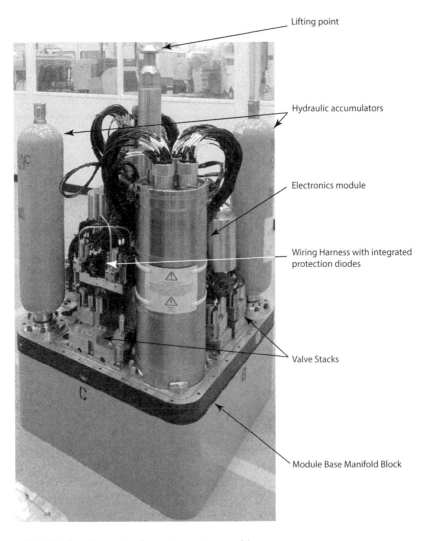

Figure 14.5 Hydraulic control equipment assembly.

the manufacturer's factory. When the tests were performed on the first units that came off the production line, some of the tests revealed failure conditions often associated with the bundles of black cables visible at the top of the photograph. The test engineers requested that the cables be replaced before repeating the test that had produced a failed result. However, they did not return to the start of the test sequence: they simply repeated the failed test. If the unit passed, then they continued with the remaining parts of the prescribed test sequence. Many of these units subsequently failed within a few weeks of being installed at the end-user's facilities, even though they were designed to have a service life of at least 30 years. The procurement engineers working for the project management organisation forgot to include a requirement to repeat all the tests after making any repairs to rectify failures detected during the factory acceptance testing. The cause of the failures was unreliable cable connections. When the cables were replaced, new faults were introduced that were not detectable unless the entire testing sequence was started from the beginning.

Responsibility for inspections and testing

Every supplier has an interest in ensuring that its components and materials pass the specified tests and inspections. If these tests and inspections are performed by staff from the supplying organisation, there will be a tendency for them to portray the results in the most positive way possible.

In the same way, engineers working for the procurement or project management organisation want to avoid, at all costs, a situation in which they accept equipment or services, only to later find that there are defects they overlooked. Therefore, they may tend to interpret any test or inspection result in the most negative way possible, adding to the costs sustained by the supplier and possibly leading to contract disputes.

It is also possible that the engineers working for the procurement or project management organisation have received inducements from the supplier organisation to adopt a lenient or generous opinion when interpreting tests and inspection results.

For these reasons, it is not unusual for acceptance tests and inspections to be performed by a neutral third party, one without an interest in portraying the results in any particular way. There are international firms, such as TUV and Bureau Veritas, that perform independent testing and inspections.

Risk analysis and management

Project planners assess the likelihood of foreseeable but unpredictable events that can influence the project schedule or outcomes. Each is assessed as critical, moderate, or unimportant by considering the consequences; controls (or back-up plans) are created to reduce the likelihood and/or consequences of such events.

Approvals

Engineering projects can cause significant, unintended consequences for the surrounding community. Because of past failures, engineers now find themselves under detailed scrutiny by government regulatory agencies, which mostly employ engineers. Therefore, engineers find themselves devoting considerable effort to writing applications seeking approval for their projects from government agencies.

Even a small project can require 20 or 30 separate approvals, both within the project organisation for the plan and the budget and from government regulatory agencies. A large project in an industrialised country may require as many as 150 or even 200 separate approvals from several different local, regional and national government agencies.

Final Investment Decision (FID) approval

The most critical approval is the one that authorises finance for the project, beyond the initial project definition and planning.

Experience shows that projects with detailed plans prepared are much more likely to achieve their objectives. The concept of 'front-end loading' represents the degree of effort devoted to planning and preliminary design for a project in order to maximise the chance of achieving the stated objectives. Once final approval is given to proceed, work on the project activities listed in the WBS can commence; the monitoring stage of project management also commences at that point.

Monitoring progress—continuous learning

Once the project work commences, project management shifts from planning to monitoring.

As explained before, the essence of project management work is coordinating collaborative work performed by many different people, possibly in many different locations. To do this, it is essential to monitor the progress of every activity, which is not as easy as it looks.

Monitoring requires a cycle of activity, much of it composed of learning. The cycle is repeated regularly, but the time interval depends on how the project is being managed and by whom (Figure 14.6).

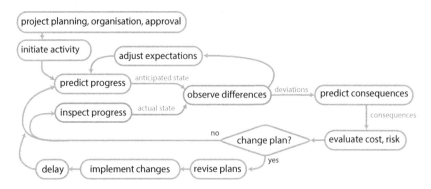

Figure 14.6 Project monitoring is a cyclical activity. It is identical in principle to the monitoring phase discussed in Chapter 12, except for the reliance on extensive information in project documents that must be acquired by the people who need it.

Many describe the monitoring process as a simple, endless, repeating cycle: plan, do, check, act, plan . . . One starts with a plan, and then people follow (do) the plan, someone checks the results against expectations, one takes any necessary corrective action and then adjusts the plan, if necessary—repeating the cycle until the project is completed. However, this can easily overlook the learning steps along the way, finding information in documents and computer systems, and discerning what is relevant.

A project engineer or manager not located at the site will typically check on progress every week. They will visit the site and spend the first hour or two with the site supervisor or site engineer, working through every activity and reviewing records of progress. Then, together, they will walk around the site looking at locations where there are hold-ups, unexpected issues, or mistakes that need to be rectified, as well as simply conversing with sub-contractors, tradespeople, and technicians.

Before each inspection visit, a project manager will predict the results from each currently running activity: what they expect to see. By comparing actual progress with expected progress, one can 'calibrate' one's expectations so that predictions become better and better over time.

Critical to all of this, of course, is establishing some way of assessing progress on each activity. Anything that results in tangible production is relatively easy to monitor. The difficulty with a lot of engineering work is that it is intellectual and cannot be directly observed. Intermediate results often give little indication of actual progress. Design work ultimately emerges in the form of drawings and documentation, but these appear relatively late in the process. In the early stages, the best one can expect to see is often just a series of sketches backed up by a conversation with the designer. Here, one can begin to appreciate how important it is to think in advance about the questions posed earlier in relation to each activity.

- How can we assess progress, quality, and other attributes of the work in question well before it is actually finished?
- How can we tell if it's going to be completed on time, at an acceptable quality, within the budget, and sufficiently long before it is actually needed so corrective action can be taken if required? And, finally——
- How can we tell that it really has been completed with sufficient quality, accuracy, and within the allocated budget?

One of the most difficult aspects of monitoring progress is a tendency for many engineers (and others) to engage in self-deception, particularly with engineering technical work that is highly intellectual and depends on abstract thinking, such as design and planning.

Many engineers tend to overestimate the proportion of work that has truly been completed. It takes someone with intimate knowledge and deep experience to ask appropriate questions that can expose the gaps that engineers believe have been covered and resolved.

So, as we can see from this brief discussion, the monitoring phase of a project—with its repeated prediction, inspection, evaluation, and review steps—is time-consuming. A project manager's job is not easy. Also, it can be difficult for others to appreciate the amount of time and effort involved. The project management literature—while often neglecting issues such as the effort required to learn the content of relevant documents—describes many different methods that can be applied to assess overall project progress.

Checking that each activity has been properly completed is also critical for financial management of the project. Project managers authorise payments for contractors once they have completed agreed-upon work on the project. If a payment has been made and, subsequently, mistakes are discovered, it may be difficult or impossible to persuade the contractor to rectify the mistakes, even if there are warranty clauses in the relevant contracts.

Completing the project

A project is completed when all the agreed-upon objectives have been achieved and accepted by the client. Completion is relatively straightforward if the project definition phase included all the necessary details on how the completion of every task was to be certified with appropriate acceptance tests and inspections. Inspection and testing activities would normally be included as an activity in the WBS. Therefore, completion of the project corresponds with the completion of all the activities.

In reality, most projects end up with substantial lists of items that need to be inspected or rectified. The benefits of detailed documentation at the start of the project now become apparent: without details, a project manager can easily end up in protracted negotiation with the client on how to complete each activity. Naturally, the project manager wants to keep the overall expenditure within the budget or as close to the budget as possible (Figure 14.7). The client, on the other hand, has a vested interest in minimising payments to the contractor (or the project organisation).

Normally, the project manager is under intense pressure to deliver the agreed-upon objectives within the original project budget. Unless project documentation is sufficiently detailed, this tension presents the project manager with the tempting option to remove non-essential items from the scope or to simply ignore them, particularly if they will cause the project budget or completion timescale to be exceeded. It's not easy for a project manager to negotiate an extension to the budget or time schedule as the project draws to a close unless there are well-established reasons that were documented earlier in the project. Inevitably, therefore, many issues are simply ignored and become operating and maintenance issues that must be sorted out later, often by a different owner.

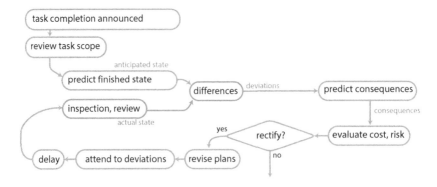

Figure 14.7 Activity task completion.

References

Hartley, S. (2009). *Project Management: Principles, Processes and Practice* (2nd ed.). Sydney, Australia: Pearson.

Merrow, E. W. (2011). *Industrial Megaprojects: Concepts, Strategies, and Practices for Success.* Hoboken, NJ: John Wiley & Sons.

Project Management Institute. (2017). *A Guide to the Project Management Body of Knowledge* (6th ed.). Newtown Square, PA: Project Management Institute, Inc.

Twort, A. C., & Rees, J. G. (2004). *Civil Engineering Project Management* (4th ed.). Oxford: Elsevier Butterworth Heinmann.

Whyte, J., & Lobo, S. (2010). Coordination and control in project-based work: Digital objects and infrastructures for delivery. *Construction Management and Economics*, 28(6), 557–567. doi:10.1080/01446193.2010.486838

Winch, G. M., & Kelsey, J. (2005). What do construction project planners do? *International Journal of Project Management*, 23, 141–149.

Chapter 15

Generating value

This chapter may be new for your supervisors, mentors, and even senior engineers in your enterprise. It came from our most recent research findings: most engineers today find it hard to describe the commercial value created by their work, beyond simple efficiencies.

Understanding how value is generated through engineering work will help you prioritise your work in ways that are more likely to align with commercial objectives and build social value in the communities you serve. It will also be easier for you to gain support for your ideas from senior decision-makers, and your career will be more rewarding.

A young civil engineer I met recently was working on a major expansion project for his employer, a large multinational corporation. He felt frustrated because he was not doing any 'real engineering work' and was thinking of giving up engineering.

I asked him what he was doing.

He said,

> They just send me out to check whether contractors have installed electrical junction boxes in the right places or put the right culverts under the roads where they were supposed to. When it rains, I have to go and watch to see if the water flows through okay and take photos. Then, I have to fill out a stack of paperwork back at the office. It's not engineering work at all, definitely not what I expected. I'm also not learning anything that's going to help me in my engineering career.

I explained that the work he described was engineering work. In engineering, it is vital to check that the contractors have completed all their assigned work in compliance with the requirements in the contract documents before they're paid.

I said, "It's just that your engineering lecturers didn't tell you about many important elements of engineering work, elements that are often just as important as performing structural design calculations."

Once the contractors have been paid, it can be very difficult to get them to come back and fix mistakes without additional payment. It can be even more expensive to pay someone else to fix the mistakes, and it usually requires high-level engineering knowledge to spot those errors.

This young engineer was contributing value by reducing the chances that mistakes had been missed, as well as avoiding the risk that extra money would be needed to fix them. He saved hundreds of thousands of dollars in just a few days by spotting mistakes.

In this chapter, I explain how engineers generate value for their firms and clients. The ideas in this chapter arose very recently in our research: you may not find them anywhere else right now. It was somewhat surprising to find that there is so little written about how value is generated from the contributions of engineers.

Engineers generate value in three ways.

Value creation: engineers plan projects and create sufficient confidence that investors will provide money for projects. The value created is equivalent to the amount the investors provide.

Value delivery: using money from investors, engineers convert plans for projects into real objects, systems and services that provide value to end-users who pay to use them. In other words, engineers deliver the value that was potentially created when the project was planned.

Value protection: engineers protect value represented by objects, systems and services, which would be lost without the efforts of engineers that sustain them. Energy, water, transport, communication and sanitation services are critical for the functioning of all human societies: inadvertent failure can lead to enormous value destruction, disease and loss of life, far beyond the replacement value of the engineered systems themselves.

A productive enterprise generates more commercial value than its operating costs. Value, in this chapter, means much more than money, even though most engineering is constrained by the availability of finance. Value is a subjective perception, and has been described as having two distinct 'flavours':

i   Exchange-value: an amount of money that a buyer is prepared to exchange for an object or experience, and
ii   Use-value: perception of an experience, or the feeling of entitlement to an object.

Therefore, given that private and government investors provide most of the finance for engineering work, much of the value generated by engineering work arises from the ways that engineers influence perceptions in the minds of investors, as well as end-users.

Furthermore, value generation must be seen more broadly than the economics of the enterprise alone. An enterprise can only flourish because of the community that supports it; therefore, an enterprise must also generate social and economic value for the community.

It's a good idea to understand value generation. Here are three good reasons:

a   Rewards
    Research shows that engineers who learn how value is generated through engineering tend to earn higher salaries, as their employers see them as being more valuable for the firm.
b   Reputation
    Engineers who understand value generation know that the seemingly non-technical, human aspects of their work, and mundane jobs like checking and inspections are critical for delivering commercial value and protecting existing social and

economic value. They will be more likely to exceed investors' and users' expectations for on-time delivery and service quality—factors that build reputation.

c   Security

Engineers who can explain how they generate value are less likely to be laid off when business conditions deteriorate because they can more easily explain why they should remain employed by the firm in ways that make sense for decision-makers.

It may take time for you to appreciate that 'value' is mostly a subjective perception that drives human decisions. Most of the time it cannot be measured and is often hard to quantify until after decisions have been made. It can be easy to lose sight of these ideas when you're working through the details of a project on a fixed budget. However, value perceptions ultimately drive the decisions that provide the finance you need to work as an engineer; for that reason alone, this chapter is essential reading.

Here are some of the ways that engineers create, deliver, and protect value. Figures 15.1 and 15.2 illustrate how these performances help to influence investors' perceptions that, in turn, drive investment decisions that provide the money needed to do engineering work.

Innovation, research and development (1)

Engineers create value through innovation, research, development, experimentation, and intellectual property such as patents and designs. Most innovation creates improvements to existing products. Some lead to completely new products or services

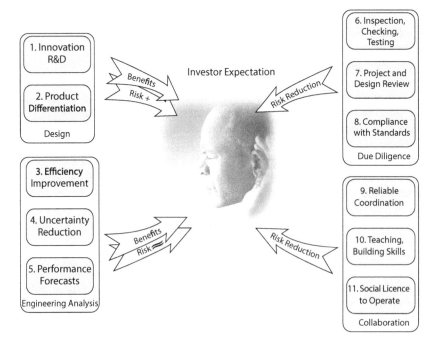

Figure 15.1 Influencing investor decisions. ('risk+' indicates additional risk perception; 'risk≈' indicates risk assessment.)

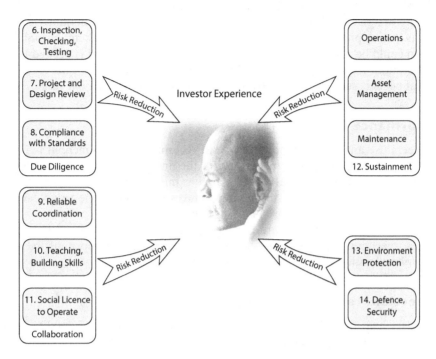

Figure 15.2 Influencing investor's experience and protecting value. These aspects of engineering help to ensure that investors, end-users, and communities see the promised benefits.

that create new markets and business opportunities. The value can be quantified only when investors contribute money to turn ideas into reality.

Product differentiation (2)

Engineers create value by designing products that provide improved buyer and end-user experience (product differentiation), since people will pay more for these. For example, an attractively designed and packaged product can improve the appearance of the retail outlet in which it is displayed, adding value for both the retailer and product manufacturer. The value created can be evaluated from increases in prices paid and sales volume.

Efficiency improvements (3)

Engineers create value by minimising human effort, material usage, energy consumption, health risks, and environmental disturbance, reducing the cost of engineering work.

Reducing technical uncertainties (4)

Engineers use extensive analysis, calculations, and experiments to reduce technical uncertainties. By doing so, they reduce the additional human effort, materials, energy, and environmental disturbance needed to ensure a given outcome with a given probability of success. These additional provisions are often referred to as a 'design margin', 'design factor' or 'safety factor'. Consider, for example, an improved material manufacturing process that provides a material with reduced defects. A smaller design margin could be adopted by using this material, as less material is needed to provide the minimum required strength. Reduced material consumption then leads to commercial benefits such as reduced transport costs, creating additional value.

Performance forecasts (5)

Engineers use analysis methods to forecast technical and commercial performance with sufficient apparent certainty to justify the financial investment. Engineers also use analysis methods to identify the relative significance of risks and uncertainties, thus reducing uncertainty in the minds of investors. Again, the value can only be quantified (if at all) after a subsequent investment decision.

Much of the time, engineers are expected to provide forecasts with quantified uncertainties to account for significant data gaps. Remaining data gaps must be identified and replaced with justifiable assumptions.

Inspection, testing, and design checking (6)

Engineers help to deliver value with inspections, checking, and quality assurance systems. External auditing can be critical for reassuring investors and reducing their perceived risks. The value delivered can only be appreciated once the project is successfully completed, by comparing it with projects that fail due to inadequate checking and quality assurance.

Project and design reviews (7)

Engineers help deliver value by arranging external reviews by experts who examine project plans and designs. External experts can often spot potentially expensive omissions. The reviews will often recommend that engineers perform additional work, such as checking the accuracy of earlier predictions. These checks are in addition to internal reviews and formal document checking required for quality assurance. Again, the value can only be quantified (if at all) after a subsequent investment decision.

Compliance with standards (8)

Another way that engineers deliver value and reduce uncertainty is by following technical standards. Technical standards guide engineers towards error-free solutions more quickly. For investors, this also reduces risks, helping to create greater value. Again, the value can only be quantified (if at all) after a subsequent investment decision.

Reliable technical coordination (9)

Engineers also deliver value by aligning the collective actions of many different people with the original technical intentions, reducing misunderstandings and misinterpretations. They do this sufficiently well to achieve the expected level of technical and commercial performance, within time and resource constraints.

With so many people involved, there are many uncertainties in the delivery of large engineering projects.

Therefore, reliable technical collaboration not only creates value by reassuring investors before they make the final investment decision but also helps to deliver the anticipated value represented by that investment decision. Evidence for this comes from industry evaluations of large engineering projects where, for example, typical projects fail to meet completion expectations two-thirds of the time, providing less than 50% of investors' expected financial returns. Therefore, it is not unreasonable to quantify the gains from effective collaboration at, potentially, about 30%–50% of the investment value of a project. A project that fails to deliver investors' expectations not only destroys value but also damages the reputations of all involved.

The value delivered can only be appreciated once the project is successfully completed by comparing it with projects that fail because of collaboration failures, one of the most common causes of project failures.

Teaching, building skills (10)

Engineers help deliver value through teaching and building skills. Most large project failures stem from human misunderstandings or misinterpretations that can be avoided with appropriate teaching by engineers. Investors value a project more highly if they can see high levels of skill in the enterprise.

Social licence to operate: co-creating value with communities (11)

Engineers help deliver and protect value with comprehensive safety and environmental monitoring practices. In essence, these engineers are creating and maintaining a 'social licence to operate': a high level of trust from the local community and government regulators. Without such trust, a company will either encounter significant regulatory obstacles or, worse, face the prospect of being closed down in response to what may be ill-informed community protests. Of course, engineers are also helping to maintain safety and protect the environment. A social licence helps investors value a project or enterprise more highly.

Engineers who engage with the local community, build skills, and even empower people to take an active decision-making and monitoring role are co-creating value. Long-term success for an enterprise depends as much on nurturing the community that hosts it as it does on commercial performance.

Sustainment: operations, asset management, and maintenance (12)

Engineering operations, engineering asset management, and maintenance engineering (collectively known as sustainment) are critical for protecting the value embodied in engineered products, systems, and business processes. These require elaborate technical coordination and other collaboration performances by engineers. For example, a gas pipeline needs carefully planned and implemented inspections and maintenance. Without these measures, the condition of the pipeline can deteriorate, resulting in considerable value destruction.

Environmental protection (13)

Engineers protect naturally endowed value by conserving the renewable and non-renewable resources of our planet—our home—and by minimising other environmental impacts. These performances also protect value represented by a social licence to operate.

Defence and security (14)

Engineers provide many products and services that limit or prevent destructive behaviour by other people, thus protecting accumulated value represented by our society and its various cultures and civilisations. Value is created even if a conflict never occurs. First, defence systems have a deterrent value, reducing the likelihood of destruction caused by actual conflict. Second, in the event of actual conflict, good defence equipment limits destructive behaviour and reduces the extent of destruction sustained.

Small and medium enterprises

Let's see how value can be generated in the context of a small engineering company that manufactures and installs electronic fire alarm systems in buildings. Electronic smoke detectors and other fire sensors are all wired to a fire alarm control system panel. A microcomputer in the panel responds to a smoke alarm and causes a fire alarm to sound as a warning to people in the building. It also sends a signal to the local fire brigade, automatically providing the address and directions so help is able to reach the building quickly. The panel contains a public address system with a microphone, enabling the fire brigade to make announcements throughout the building when they arrive. The panel provides internal wiring connectors for the smoke detector sensors, a public address amplifier, switching circuits, and communications circuits.

The company manufactures the panels using locally sourced and imported components. The owner started the company by utilising his practical electrical wiring and circuit board assembly skills. Other members of his family work for the company in administrative and accounting positions (Figure 15.3).

Here are some ways that engineers can generate value for the company.

Product and process improvement, research and development, and anticipating future developments

Engineers prepare designs for an upgraded or lower-cost product that could be manufactured in the event that a competitor enters the market with a better-quality product at a similar price or a similar product at a cheaper price and evaluate the likelihood that this might happen. They generate value by reducing the risk that a competing product will take a significant portion of the existing market share.

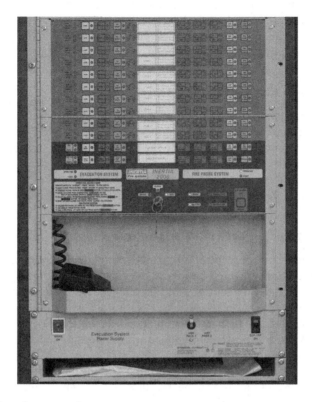

Figure 15.3 Fire alarm panel.

Collaboration

Engineers improve collaboration and technologies in the company to provide better service quality. They generate value because improved client satisfaction and customer loyalty help to grow the business.

Business development research and understanding customer needs

Engineers work with fire insurance companies and demonstrate the company's product compliance with international and commercial fire safety standards. With appropriate certifications, the company can generate value by opening new market opportunities because certified fire protection equipment can reduce customers' insurance premiums.

Cost monitoring, control, and reduction

Engineers investigate and, if necessary, help to rearrange the company's accounting systems to ensure that the different costs of manufacturing electronic panels, the costs involved in servicing and maintaining products installed at customer sites, and the costs of maintaining or replacing production and service facilities and equipment can be accurately monitored. They generate value by accurately monitoring costs, enabling the company to identify where technical improvements could provide additional value, and demonstrating which previous improvements have provided real benefits in order to improve cost-efficiency.

Risk management and reducing uncertainties

By demonstrating compliance with international standards for all aspects of the product, as well as company operations, engineers can increase the likelihood that future company expansion will be financed by the company's bank (rather than family capital) with a lower cost of capital. Banks closely inspect a company's operations when a company requests bank financing, often employing engineers to perform these investigations. This work generates value by reducing the cost of finance.

Balancing value generation with cost

I have explained how perceptions of value motivate investors (and end-users) to provide finance for engineered products, services, and information. Engineers can strongly influence these perceptions and thereby generate value, which translates into a willingness to pay.

In the end, of course, a firm must be financially sustainable; it must influence investor perceptions to attract sufficient finance and earn enough income from end-users to cover the cost of running the business.

Therefore, engineers must be able to predict and ultimately control expenses to ensure that investor and end-user expectations are met. Therefore, the next important aspect to learn is cost estimation.

Quantifying value generation

First, recall that value, in essence, is a subjective perception, and everyone has a different perception that depends on time, circumstances, and context. Therefore, most aspects of value cannot be quantified readily until, as a result of subjective perceptions, a sponsor or investor makes funding and tangible resources available in response to those perceptions, or parties agree on an exchange-value. Therefore, how the value is portrayed to the sponsor, buyer, or investor is as important as generating value in the first place. This can only be assessed through conversations with people, listening carefully to indications of their attitudes and perceptions, and noting subtle changes in word choice as their perceptions change.

Risk, in particular, is one such perception. Notice how rare it is for a passenger not to fasten their seatbelt in an aircraft, compared with riding in the backseat of a car, for example, even though the risk of accident in a car is far higher than when riding in an aircraft. Quantitative risk and perceived risk are completely different in identical circumstances.

Of course, performance forecasts in engineering must be quantified. However, perceptions of value may influence the reader's assumptions, thereby influencing the reader's interpretation of those forecasts. Trusting personal relationships also influence value perceptions; indeed, without them, most investments would never happen.

It is no accident that financial centres tend to be in cities with office and residential towers that enable a large number of people to meet each other and spend time building the trusting relationships upon which financial investments depend, all involving differing perceptions of value.

To keep it simple, therefore, forecasts are expected to be quantitative, bracketed by semi-quantifiable expressions of likelihood. Value perceptions strongly influence the way those forecasts will be interpreted, and ultimately, that is what determines the level of investment or the exchange-value.

Learning more

Reading financial news will help you appreciate which industries and companies are creating more commercial value in your country, and therefore could offer attractive career prospects. You will learn some of the language of finance by doing this as well.

Reading the annual reports of companies operating in your industry will also help you appreciate how commercial value is described and generated, and this will serve as a basic introduction to accounting. Reading a basic accounting text will help you understand more about annual reports.

Chapter 16

Estimating costs

Engineering is expensive. Engineers can quickly gain respect for accurate cost estimation and the ability to perform and coordinate work effectively within allocated budget and time allowances. Learn from an experienced estimator in your industry as soon as you can.

While estimating is an engineering function, setting the price is a sales and marketing issue. The price of a product or solution often reflects an assessment of customers' willingness to pay: the exchange-value. Therefore, the price may be quite different from the estimated cost; the price should normally be higher than the cost for financial success. Sometimes, however, the firm will accept a lower price to secure the work in the belief that doing so will lead to later, more profitable opportunities.

Estimating

The starting point for estimating is a full set of documents: specifications, functional diagrams, and detailed drawings. The estimator uses these to create a complete Bill of Quantities (BoQ, equivalent to Bill of Materials, BoM), 'taking off' all relevant details such as quantities, sizes, lengths, areas, volumes, and masses. He or she will have access to standard 'rates' taken from past experience or industry reference texts to calculate the total cost of each item, including manufacture, storage, transport, installation, and acceptance testing. For example, here are sample rates for a certain kind of pipe:

- 0.21 work hours for cutting to length,
- 0.16 work hours for bevelling each end,
- 1.5 work hours for welding by a skilled welder,
- 0.82 work hours to install per metre of length, and
- 2 pipe supports per metre.

This excludes painting, finishing, and insulating the pipe. Another reference source might simply tell you that, in a process plant, the same pipe requires 0.82 work-hours per metre of length. Experience will guide an estimator in choosing which rates will be appropriate. In addition, there may be 'uplift' or 'multiplier' factors, usually greater than 1, to account for hazardous conditions, or installing the pipe in difficult to reach or elevated locations. A multiplier of 1.2 might be used to account for material handling and transporting cranes, tools, and people on a very large site. Multipliers of 3–5 apply for remote sites and 8–10 for work offshore, on top of a minimum USD 30,000 to organise even a small project in such locations.

Often, however, a large proportion of the work cannot be estimated this way. The drawings may not yet be completed, perhaps the design itself is incomplete, or only available as a tentative outline, adding considerable uncertainty. A rough estimate could be based on the capacity of a process plant, the number of lanes, intersections and length of a highway, the antenna size and transmitting power of a radar, or the range of operator functions, data types and data handling capacity required for some software.

Much of the estimation may have to be performed by subcontractors who will likely perform aspects of the work if the firm wins the contract. Experienced engineers know how reliable these estimates will be, given past performance by different contractors.

Estimators provide estimates for 'risk factors' to cover relevant uncertainties. The relative magnitude of risk then guides negotiations on how the project should be handled. Here are four possible ways, which might be used in combination:

1 Fixed price: an experienced engineer or firm takes the risk factors and estimates into account and provides a fixed price to perform the work;
2 Schedule of rates: the client agrees to pay the firm for the work based on agreed-upon rates for all work still to be defined;
3 Cost-plus: the client agrees to pay all the firm's costs, plus an agreed profit margin;
4 Alliance: the firm and the client effectively merge into a single organisation for the project duration and share all normally confidential commercial information, with an agreed-upon arrangement for sharing commercial benefits from the project.

In most instances, a client will seek 'tenders' from several firms for a given project: detailed written submissions on price and schedule, explaining how the work will be performed to the required quality level. Therefore, to be awarded one contract, most engineering firms must submit between four and ten tenders with cost estimates. Frequently, the contract will not be awarded without extended negotiations, with the client asking for estimates to be re-done several times for different options. Naturally, there is nearly always a tight time schedule, with as little as 2 weeks to complete estimates on complex designs. Coordinating estimation work across many subcontractors and separate design groups can be the most challenging aspect for preparing tenders.

Schedule, quality, scope, and commercial considerations all influence pricing. If the client specifies a 'tight' schedule requiring very careful planning to avoid delays, there will be cost implications. Specialised work may need to be performed in parallel, requiring two or more teams and special equipment such as cranes on a construction site. Multiple shifts[1] may also be needed so that work can proceed through the night.

The scope may be negotiable—for example, the client may accept a lower performance specification for costly equipment. Exploring the client's real needs through discussion, and possibly working with end-users or the client's customers, might lead

---

1 Work is arranged in shifts when people are required for more than a normal span of working hours. For example, 24-hour site work may be arranged in two shifts of 12 hours or three shifts of 8 hours.

to considerable cost savings by excluding unnecessary parts of the project work scope. Discussing the client's needs with equipment suppliers can also help reduce costs: the equipment suppliers can often explain how other customers have changed their specifications to secure large cost savings.

Maintaining high-quality standards may be critical—for example, even the paint-work has to be rigorously inspected on an offshore gas processing facility to guard against premature corrosion that could weaken highly stressed pipework and structures. Close supervision and necessary inspections increase costs.

Perth engineer Peter Meurs managed to complete a large mineral processing plant at an iron ore mine for two-thirds the estimated cost and completed the work in a shorter time than planned, by arranging for meetings between the project owner and major equipment suppliers at their factories. By understanding the project owner's needs first-hand, the equipment suppliers were able to suggest many cost- and time-saving options of which the plant design engineers were unaware.

Labour cost

One of the most frequently misunderstood estimating factors is the real cost of labour.

It is easy to focus on the direct salary or wage cost of the labour: what a person gets paid every week, month, or year. What matters more is the total cost to achieve a given result with appropriate workmanship. The cost of land, machinery, energy, and materials needed for a product or project can be more significant factors. Low standards of workmanship can result in wasted material costing far more than the labour.

The cost of employing a person is much greater than their wages (or salary), the 'direct' cost. There are also many 'indirect' costs. (Since terms such as direct and indirect costs are used in different ways, it's best to ask what they mean in your situation.)

There are standard 'on-costs', usually calculated as a percentage of the wage or salary, with rates depending on local regulations.

- Annual leave and public holidays: 6%–9%. Most workers are paid when they take their leave, so there is an additional cost in proportion to the amount of time taken as leave;
- Sickness and caregiver leave: 2%–3%;
- Employer liability insurance: 3%–10%. Compensates workers for industrial accidents;
- Pension contribution: 3%–10%;
- Tax on salaries: 3%–5%

Total on-costs are typically 25%–35% of total pay.

There are several indirect costs as well.

- Administration, accounting, and compliance with employment regulations: ~10%;
- Accommodation (or rented space), energy for lighting, heating, air conditioning: ~10%;
- Supervision: 10%–35%. The job of a supervisor can be very complex and requires substantial experience; in a less developed country, the cost of supervision may be 300%–400% of salary;

- Uniforms, personal protective equipment: uniforms can convey a sense of pride and prestige that can boost productivity;
- Special tools, workstations, vehicles, training courses, transport, away-from-home accommodation and food, non-productive time

If the work is being performed in an environment where there is social or political instability, the cost of providing security can be one of the largest components of the cost of labour. If the need to provide security interferes with the work itself, then lost production time is an additional cost.

Part of the time spent at work will be needed for safety briefings, transport between work locations on-site, rest breaks, eating, refreshment, and toilet breaks. This is often labelled as 'non-productive time' (as opposed to 'time on tools'). It is easy to see socialising during breaks and around a coffee machine as non-productive time. However, as we have seen and discussed, engineering enterprises rely on distributed knowledge and collaboration; the knowledge required for the enterprise is carried and transformed by the minds of the people who work for it. Social contact is critical for this. Therefore, time spent socialising is a necessary component of working.

A normal 38-hour workweek provides almost 2,000 working hours per year. Of this, one can expect around 1,500 productive working hours in a year (125 hours per month, or about 29 hours per week).

Next, the labour estimate is adjusted by multipliers to cover project-specific factors that influence individual worker productivity. For example, a project that requires two or more shifts will normally require some workers, particularly supervisors, to be present for handover briefings for people on the following shift. Time-critical tasks may require additional labour to be present to ensure that they are done on time. Some work may also require time spent on cleaning up the workplace.

Further factors are applied to account for interruptions caused by, for example, wet weather on a construction site.

In a low-income country, local workers are usually much less productive than skilled workers brought in from an industrialised country because they have had less training. More working hours will, therefore, be required.

The cost of labour, including indirect components, is usually only a small component of an engineering enterprise. The labour cost proportion ranges from around 8% in a low-income country to 15% or 20% in an industrialised country. Capital costs—equipment, depreciation, land, and consumables such as materials, energy, transport, and financing costs—make up the rest. Labour productivity, therefore, rests on the ability to make effective use of these other resources.

Many people perceive labour costs in terms of the most obvious direct wages or salary component. This misunderstanding can have serious consequences. It is more important to think about productivity than hourly labour costs.

For example, the total cost of using a large earthmoving machine can be $150–200 per hour, including fuel and other consumables, maintenance, support costs, and leasing payments. Machine productivity depends much more on the skill of the driver than on what the driver gets paid. Part of the driver's skill is the ability to drive the machine carefully to minimise non-productive time for machine maintenance: the fastest driver is not necessarily the most productive. Many companies using large and expensive machines hire women to drive them because they are more gentle than male drivers, saving downtime and maintenance costs.

As with the capital expenses, there are many other expense categories not mentioned here (see the life cycle costing guide in the online appendix on https://www.routledge.com/9780367651817 for a list of other categories). Experience plays an important part as well; experienced engineers develop an understanding of how much labour is required to achieve a given result in a given situation. Estimating is a special skill and expert estimators are vital to any engineering enterprise.

What does it cost to employ you?

The cost of employing engineers is also significant for an engineering enterprise.

Naturally, the exact costs depend on the particular company and its respective circumstances. The figures below indicate the approximate costs in Australia at the time of writing but will vary from place to place. However, in most settings, the costs will be more or less similar in proportion.

Obviously, the first component is the engineer's salary, between AUD 6,000 and 12,000 a month at the time of writing (USD 4,000 to 8,000 at current exchange rates).

Other costs include the ones listed in Table 16.1, expressed as an approximate percentage of salary.

The last item (business development) represents the proportion of an engineer's time that will be needed to support business development: winning contracts for the firm, helping to write proposals, and tendering submissions to gain new work. The most senior engineers may spend up to 40% of their time on business development.

With 4 weeks of annual leave and a nominal 38-hour workweek, an engineer will be at work for a maximum of 1,800 hours each year. Allowing for 1 week of sickness, time for personal hygiene, work breaks, and moving from one location to another (200 hours), business development activities (300 hours), and training and guidance from supervisors (300 hours), approximately 1,000 hours are available for 'billable'

Table 16.1 Typical overhead costs for engineers

Payroll tax, employer's liability insurance	5%
Superannuation/pension contribution	10%
Annual leave	8%
Office accommodation rent (approximately)	15%
Administrative support	15%
Supervision and assistance from senior engineering staff, approximately one hour per day for the first year and about 30 minutes per day thereafter	33%
Health insurance contribution	2%
Electricity, water	3%
Travel, accommodation, visits to engineering sites, attendance at conferences, sales seminars, etc.	15%
Engineering software licence fees	10%
Subscriptions to databases, libraries of standards, and other sources of information	5%
Business development, preparing proposals and contributing to presentations by other company staff	25%
Total	146%

work, which are projects for which a client will pay. In practice, it is difficult to achieve this because there are times when there is insufficient work to keep everyone occupied. Let's optimistically assume that a young engineer would be assigned to work on projects for 900 hours annually. That means that the cost of employment must be recovered on 900 hours of work for which the firm is paid by the clients.

With a monthly salary of $6,000 ($36.50 per hour pay, before tax is deducted), the hourly rate needed to cover salary and all the overhead costs listed above will be at least $200. After allowing for a profit margin, the cost to a client will be about $220 per hour.

As a novice engineer, you need to understand that you must create sufficient value through your work to justify the cost that a client will be asked to pay for your services.

Most graduates and many engineers have never seen this simple calculation, and therefore find it difficult to understand why their managers might be concerned about the way they spend their time.

It's a good idea to be confident that you are creating at least as much value as it costs a client to obtain your services. You also need to charge clients a fee that covers your costs and a reasonable profit margin. It is easy to underestimate both of these costs.

Low-income countries

Special cost estimation methods are required for work in low-income countries. Although it may surprise you, most engineering work in these countries is considerably more expensive than in advanced countries such as Australia and the USA. The perception that labour is cheap in these countries is quite misleading. The next chapter explains why and provides details on factors that cause these higher costs.

Chapter 17

Navigating social culture

While this chapter focuses on low-income countries, social culture shapes collaboration everywhere. Engineering practice consists of two distinct threads. The first uses engineering science knowledge for analysis and performance predictions. The principles and methods are similar everywhere. The second and much more time-consuming thread is organising effective technical collaboration: ensuring that the actions of everyone involved will preserve original technical intentions so that ultimate performance aligns sufficiently with expectations. Collaboration can only be sustained by navigating local social culture.

What is culture? In essence, it is the habitual ways that people interact with each other. Every country, every region within a country, every community within a region, every company: each has its own culture, its way of 'doing things with others.' Often one is more aware of culture differences as a stranger, though of course, it takes time before you can navigate the culture like a local.

I have learned through 20 years of first-hand experience and on-ground research in India, Pakistan, China, and several other countries, that it is more difficult for engineers in low-income countries to build the same level of trust, knowledge sharing, and collaboration as they can in a wealthy country such as my home in Australia. Also, it is much more difficult to learn from specialised engineering product suppliers, as there are so few knowledgeable sales agents in these countries. Furthermore, there are several persistent misconceptions that tend to influence decision-making inappropriately. Therefore, most engineers in these countries tend to be much less productive, and they earn much less because they are less productive. Their enterprises are mostly less productive as well. That means the cost to produce equivalent goods is higher. For example, electricity at the point where it performs useful work is typically 3–5 times more expensive in low-income countries.

This is not because engineers in wealthy countries are cleverer, wiser, better educated, or work for rich companies.

During my research, I was surprised to find a tiny number of engineers in India and Pakistan, earning more than their counterparts in Europe and America. Gradually, I realised that these 'expert' engineers had learned for themselves to be much more productive than other engineers. Their firms could not afford to lose them. Therefore, they were paid very high salaries, high enough to dissuade them from leaving to find work in a wealthy country. Their self-taught knowledge enabled them to navigate the complexities of South Asian social culture. Much of what they had learned for themselves is found in this and other chapters of this book, as well as more extensively in my earlier book, *The Making of an Expert Engineer.*

Why is this discovery so important?

It shows that it is possible to become a highly productive engineer in a low-income country and earn a salary higher than in a wealthy country. In this chapter, I explain how you can develop similar abilities. If most engineers in low-income countries could master these abilities, I believe these countries could be much more prosperous, sustainable, and attractive places to live.

Learning to organise effective collaboration can be challenging—years of formal education taught you to value individual work rather than collaborating with other people. The specialised engineering collaboration methods described in this book are seldom mentioned in engineering schools. In this chapter, you can learn to overcome the even greater challenges of organising effective collaboration in low-income societies.

You may be reading this chapter as an engineer who has grown up in a low-income country. You may have grown up in a wealthy country, and now you have been assigned to an engineering project in a low-income country. You may work in a wealthy country and be responsible for outsourced engineering work performed in a low-income country. Whichever the case, this chapter has important guidance for you.

What's different?

Research has revealed (so far) eight significant differences between engineering in low-income societies and engineering in wealthy societies like Australia, the USA, Europe, etc. The differences are not always what they seem to be at first glance, and some can seem counterintuitive.

For example, many workers in low-income countries seem lazy and disinclined to work. They may do nothing if they think a supervisor is not watching them. They may take shortcuts or improvise with low-cost, but unreliable repairs. Sometimes they can seem frustratingly stupid. However, viewed differently, given the social culture in which they live and work, these can also be seen as intelligent responses. Let me explain with the first difference.

(1) Respect for authority

Engineers in most low-income societies occupy a privileged niche, but one that is below the top of the social hierarchy. Nearly everyone maintains subservience to the social hierarchy, which is reflected in the firm's organisational structure. If you are a foreigner, you may find yourself near the top of the hierarchy, though sometimes it may seem you're at the bottom.

Nearly every local must also live within a large personal network of extended family and relatives, friends, and acquaintances. Together, the hierarchy and one's family and personal network provide the only social safety net: protection from personal catastrophe and the loss of earning capacity. Getting a job almost invariably requires access to someone with influence, sometimes with a significant payment. Most people are only a short step away from destitution. Therefore, loyalty to the social hierarchy and one's personal network takes precedence over everything else, including productive work.

In most low-income societies, one must pay due respect to people with authority. Often, for example, this means listening to seniors without interrupting or asking questions. It may be considered impolite to ask questions, even to clarify something. Asking can even imply that you were not paying proper attention when a senior was speaking.

It can also be considered impolite to ask a question that would influence someone to say something that would result in embarrassment or a 'loss of face.'

Here's a common situation we observed, showing how cultural inhibitions like these result in workplace difficulties.

Even a young South Asian engineer acting as a production supervisor is considered to have high social status relative to production workers. He meets with his manager, who explains what needs to be done during the day, and then he goes to brief the workers. No questions are asked.

My teaching helped me realise that most students learn very little in lectures unless they have been taught to take notes and ask questions to clarify their understanding. Therefore, in such a situation when the senior speaks for 15–20 minutes with no questions asked for clarification, it is almost certain that there are many potential misunderstandings between the manager and the young production supervisor, and even more between him and the production workers. The workers, acutely aware of unresolved uncertainties, are smart enough to know that the consequences of doing nothing are less severe than doing the wrong thing. Therefore, they patiently wait for direction from the supervisor, and expensive machinery lies idle in the meantime. The engineer may have to run from one worker to the next, explaining every small action to each of them in turn. It can be intelligent for workers to slow down or stop work altogether in the absence of visible supervision; it's sensible to wait for someone else to arrive and take the blame for mistakes. The next morning, when his manager asks for a progress report, the young engineer remains silent instead of reporting production shortfalls. The manager knows that insisting on a progress report will result in loss of face, so he too remains silent and moves on.

There are several ways to overcome these difficulties, and all require persistence.

i   Teaching. Engineers can gradually teach workers to ask questions and contribute suggestions by building their trust and confidence that these are valued contributions, not a potential embarrassment. This requires that engineers listen to workers in their workplaces to pick up their language, the words they use in their normal conversations. Trust-building depends on physical presence and extended face-to-face conversations often unrelated to workplace issues.

ii Encouraging workers to explain what they think needs to be done, in detail, step by step, as a way of detecting potential misunderstandings. Again, this takes time and careful listening by the engineer.

iii Minimising worker turnover. There are still many firms that hire 'day labour': low-paid workers who stay for a few months and then move on. This practice is reinforced by the perception that the hourly direct labour cost must be minimised. Union or government regulations may dictate that permanent workers are paid much more and are thus regarded as 'expensive.' This is a misleading perception, as explained later.

iv Finding ways for questions and suggestions to be contributed anonymously. For example, arrange for groups of workers to write questions or suggestions on pieces of paper that are placed in a pile from which one or more can be drawn at random.

(2) Navigating the labyrinth of social power

Chapter 12 explained how engineers make things happen using informal workplace relationships to gain willing and conscientious collaboration outside the lines of authority. Relying on organisation or management lines of authority causes delays and misunderstandings.

However, in low-income societies, there are many intersecting lines of social hierarchy and influence that can inhibit collaboration, even crossing lines of organisational authority.

Social class distinctions are far more important. No matter how willing, a person may not collaborate without the formal or implied consent of people with more social power. Social power can come from being a member of a powerful family, clan, tribe, or social caste. It can even come from knowledge of past wrongdoings, violations of formal regulations or informal social rules, or even rumours of such violations: memories of shame that are deeply but invisibly etched in collective memory.

It can take a lot of time and effort in casual conversations to understand this labyrinth of social influence, or just repeated phone calls, many times a day, to secure even hesitant collaboration. Without finding ways through this labyrinth and getting help from people with social power, collaborative work can take far more time and effort and is sometimes impossible.

(3) Misunderstandings on labour cost

Most people instinctively think that everything is cheaper in low-income countries. They may be thinking about the hourly pay a worker receives or remembering tourist experiences—visiting a low-income country and finding restaurant meals and taxis, for example—to be amazingly inexpensive compared with home.

In engineering, it is much more appropriate to think in terms of the total cost needed to achieve a given result with equivalent quality, reliability, and durability.

Take the cost of labour, for example.

In the previous chapter, I explained many different components of labour cost. I also pointed out that one must think of overall productivity because in many engineering enterprises, the cost of machinery, materials, energy, and land dominates the total cost.

In a low-income country, therefore, the hourly pay received by a worker is often one of the smallest cost components. Table 17.1, taken from an actual project in a low-income country around 2014, demonstrates this point. Indirect costs are much greater, and the largest single component is supervision.

Without effective supervision, very little useful work will be accomplished, as explained above.

Good supervision means much more than managing others at work and intervening when necessary to avoid mistakes. Effective supervisors plan engineering activities and see to it that tools, materials, and information are available at the right times and places to ensure that the work proceeds without interruption. They quietly train people and provide guidance, making the best use of available resources. They interpret the needs of engineers and translate them into terminology that workers understand. They use years of first-hand experience to anticipate problems early enough to avoid them.

Table 17.1 Example of hourly labour cost calculations

	Expatriate skilled trades (industrialised country)	Local skilled trades	Local labour
Direct costs			
Hourly rate with on-costs	80.00	7.00	6.00
Indirect costs			
Recruitment	4.00	0.30	0.20
Supervision	25.00	25.00	25.00
Training	2.50	1.20	1.40
Non-productive time	8.00	0.70	0.70
On-site shelter	2.00	1.00	1.00
PPE	0.80	0.80	0.80
Workshop equipment	4.00	1.00	1.00
Small tools & consumables	6.00	2.00	2.00
Light vehicles	9.00	5.00	5.00
Site office overheads (light, power, first aid, security)	2.00	2.00	2.00
Accommodation	8.00	—	—
Total indirect costs	71.30	39.00	39.10
Administration (10% of direct and indirect costs)	15.13	4.60	4.51
Total cost (USD)	166.43	50.60	49.61

Many firms in low-income countries employ graduate engineers as production supervisors, greatly underestimating the skill levels and experience needed. This is another factor contributing to low productivity.

The low labour cost perception is therefore a misunderstanding, and one that can influence many inappropriate decisions. For example, many firms in advanced countries have tried to outsource engineering work to low-income countries, only to find that the total cost is far higher than they expected; often higher than doing the work at home, when all the indirect and opportunity costs are taken into account. In low-income countries, one often finds that a firm has hired too many people, sometimes to satisfy family obligations to employ relatives. The result is a high workload for supervisors and often an overall reduction in productivity.

> Hourly labour costs for an engineering project in a low-income country (in USD) adjusted for costs in 2014. These calculations are based on research notes and do not include profit margin. Note that supervision accounts for 50% of local labour cost, while the direct rate of pay with on-cost is only 12% of the total local labour cost.

Expect the overall cost of most engineering activities to be considerably higher than in an advanced country like Australia or the USA, with similar quality and durability expectations.

(4) Documentation and organisational procedures

Engineers in advanced countries often complain about the number of organisational procedures they are required to follow. However, procedures and processes enforce collaboration, ensuring that information is exchanged between people who need it and guaranteeing that adequate records are kept. Organisational procedures make it more likely that people whose knowledge is needed to make the best decisions and avoid future problems contribute their advice at the appropriate time. Procedures embody knowledge from past experience: lessons learned about how to effectively organise collaborative technical work.

Many firms in low-income countries have weak or improvised engineering procedures or even none at all apart from approval for purchases.

Ask, for example, about configuration management. Find out whether there is a procedure in place to ensure that design changes are checked by everyone who could be affected; that the reasons for the design change (or decision not to change) are recorded systematically and are accessible for people who need that information. Ask where the history of all design changes is recorded, such that the relevant documents, specifications, and drawings can be located when needed.

Creating procedures might seem to be a simple exercise. Draw up a sequence of actions and create relevant checklists. However, that is the easiest part of building procedures in an organisation. The hard part is ensuring that everyone accepts the need for such a procedure and has sufficient motivation to adopt and follow the procedure without feeling compelled to do so. That requires their involvement in creating the procedure, negotiating the details, and determining how it is going to work. That takes considerable time and organisational effort on the part of engineers.

Creating and maintaining detailed documentation often seems like a painful administrative requirement of many procedures. However, leading engineering firms have learned how effective documentation can be in promoting productivity. Effective quality assurance, for example, relies on documented processes.

In low-income countries, much of the workforce has a limited ability to read normal engineering documents and drawings, so devising effective documentation that works for them can be challenging—and absolutely necessary.

(5) Language barriers

Engineering has its own language: common words that have completely different meanings in an engineering context. A washer, for example, denotes a flat metal ring for a mechanical engineer, but for most other people, a washer is a person or machine that washes clothes.

In many countries, people who provide manual labour in engineering firms speak a different language from the engineers. The supervisor, therefore, also has to be a language interpreter. Interpretation can be all the more challenging in an environment where languages are mixed. In India, for example, people can be using words from English, Hindi, and regional languages in the same sentence, often flipping from one language to another, with different accents, seemingly at random. Listeners have to distinguish the different languages, then the meanings, adding to uncertainties in translation.

It is not unusual to find that familiar engineering concepts like 'average' or 'probability' simply don't translate into a different language. Presented with this issue, the interpreter (typically the supervisor) needs to decide how to describe the ideas that are important for the engineer—in completely different ways and in ways that are culturally appropriate.

Of course, much of the meaning is lost in translation.

Local workers, perhaps unable to speak more than a word or two in English and perhaps with minimal education, are almost certainly not stupid. However, they may have learned that pretending to be stupid can be an effective way of avoiding responsibility and the consequences of making mistakes that arise from misunderstandings. My experience has shown that once they are appropriately motivated, local people willingly work hard and conscientiously, but reaching that point can be time-consuming and requires patience.

(6) Centralised decision-making

Procurement can be a common frustration for many engineers at firms in low-income countries and, to an extent, almost everywhere. Engineers provide requirements to a specialised division with people who organise the purchasing of supplies and materials, few or none of whom have a technical background.

Often the firm has preferred suppliers and enjoys special discounts on prices arising from a large volume of business. Therefore, the actual parts or materials may be different than the ones recommended by engineers. The procurement division will often find the lowest cost equivalent. That's their job: to minimise the cost of purchased materials and components. However, the lowest cost component often causes expensive problems when used in production.

Smart engineers, therefore, learn how to collaborate effectively with their procurement people. Knowing them in person helps, as does providing detailed specifications. For example, an engineer may know that a particular component from a specialist supplier, often considerably more expensive to purchase than the cheapest available, will provide the least lifetime cost. A smart engineer will ask the specialist supplier to provide the text of a specification that will enable the purchasing specialists to exclude lower-cost alternatives from being considered. It may be necessary to reinforce this with an acceptance testing procedure that low-cost alternatives cannot comply with.

Learning how to work with, rather than against, centralised decision-making can make life much easier, even though time and preparation may be needed along the way.

(7) Access to financial information

It is easier to know how to generate value for your firm if you have some understanding of financial constraints and commercial priorities. In low-income countries, particularly in South Asia, few engineers are trusted with any financial responsibility or detailed information unless they own the company.

Company owners are reluctant to reveal the true financial state of affairs to anyone except a few trusted staff. Many negotiate an agreement with their local tax collector to present a set of 'official' accounts—showing considerably less profit in order to minimise tax payments. Naturally, there is a fee paid for this service, in proportion to the tax reduction achieved. Other owners may not want shareholders, or even family members, to learn the true financial situation of the company, fearing that they might demand a larger share of profits.

For these reasons, many companies do not comply with recognised accounting standards. Consequently, even the company owners can be unaware of the true financial situation of their firms. For example, they may not account for liabilities such as depreciation of their assets, accrued benefits payable to their employees on retirement, or the future cost of remediating pollution of land or groundwater caused by company operations. This can result in profits being significantly overestimated.

Therefore, in order to generate real value for a company and propose a realistic business case for improvements, an engineer may need to estimate the real financial circumstances of the company. Suppliers can provide information on the prices of energy, equipment, and materials. Labour costs, salaries, and overheads can be estimated using the information in Chapter 16. As long as the firm's products are selling, if the real selling price is confidential, inquiring about the cost of competitors' products will help in estimating the firm's income.

(8) Learning from specialised engineering suppliers

Specialist engineering suppliers fulfil a critical role in engineering practice: they provide education for engineers on the vast array of specialised components, software, and materials that make engineering possible. Most graduates have encountered only a few basic materials and components in their education; the information is most likely out of date by the time they graduate.

Specialist suppliers invest in considerable education activity as a form of marketing. Their sales engineers provide training for using the components and materials from manufacturers that they represent, often with hands-on experiences. A major difficulty in low-income countries is that few such suppliers are represented. Even if they are, one or two sales engineers may have to work in a country with a population exceeding 1 billion people. That means it's almost impossible for engineers in low-income countries to benefit from this potential educational resource.

Some companies provide extensive online learning materials. It may be possible to use these materials to develop expertise in using the products. An alternative strategy is to attend trade shows whenever possible, for face-to-face meetings with supplier representatives. Many suppliers will only send representatives to a low-income country for a particular trade show.

Some products can succeed

Mobile phone systems have become commercial success stories in practically every country. Given the difficulties that affect most engineering activities, one must ask why mobile phones have succeeded, eclipsing the older landline phone systems, where utilities such as water supply and electricity have failed.

In our research, we identified mobile phones and some other products (e.g., cars, bikes) that have succeeded because implementation requires less extensive technical collaboration between engineers, technicians, and workers. Also, in the case of mobile phones, information technology systems protect the security of payments and build trust between the user, investors, and government.

Therefore, in the future, it is possible that emerging systems for providing trusted financing and electronic payments using mobile phones may enable a whole generation of new equipment designs. These systems may transform engineering in low-income countries. For example, expensive equipment could be provided under a leasing arrangement, whereby a farmer, for example, provides regular lease payments through a mobile phone. As people in low-income societies become more acutely dependent on their mobile phones, tying the mobile phone to repayments provides a high degree of security for the firm lending the finance to enable the purchase.

Think in terms of value generation

One common theme emerged from my interviews with expert engineers in India and Pakistan: they often explained their actions in terms of value generation. For example:

> I think about each decision I make: how is this going to justify the salary they pay me?

Chapter 15 explains how engineers can generate value: re-read this chapter and look for opportunities to create value in your enterprise.

Making practical improvements will require tenacity and patience; especially in a low-income society, it takes time to make perceptible changes in an enterprise. Small improvements will be easier than large ones. Once you have a reputation for making changes that lead to real improvements, it will be easier to gain support for the necessary resources to make bigger improvements.

Eventually, with a string of successes, you will be seen as a key person in the enterprise. Generating additional value for your enterprise is unlikely to lead to a pay rise by itself. You will need to ask for a salary increase. Before you do that, you need to make sure there is sufficient income to pay for it. If you do not have access to detailed financial information, as explained in (7) above, you may need to maintain your own estimates. Once you have reasonable confidence that the enterprise can afford a higher salary, you can reasonably ask to receive a fair share of the additional value created by your efforts.

If you are not rewarded with an appropriate salary rise, then you may have to look around for other opportunities. Try letting colleagues know that you are considering offers from other firms. In the end, you may need to find another job at an appropriate salary level and be prepared to leave. If the enterprise really values your contributions, they will make sure that you have a sufficiently high salary to ensure you don't think of leaving. If they fail to do this, it's probably better for you in the long run to find an enterprise that properly values your contributions.

Outsourcing

Many engineering firms and projects in wealthy countries outsource work to low-income countries in the belief that it will be less expensive. Mostly, after some time, they realise with disappointment that it is either more expensive than expected or the work is of low quality. Effective coordination requires experienced and expensive engineers and can often demand far more time than expected. Outsourcing can be effective as a temporary way to increase capacity that is too expensive to maintain permanently. It is possible to achieve economically effective outsourcing, but this usually requires patience. It takes time to find the best way to communicate and coordinate the work, and the best way to allocate work between the home country and the outsourcing location.

Opportunities

If you can successfully navigate the social culture, low-income countries present some of the most exciting opportunities because of the high costs for most engineered products and services. Equipped with an understanding of how to reduce costs, you can make the most of these opportunities and turn the local culture to your advantage.

The world needs you to do that.

With huge populations desperate for a better life, the sustainability of our world depends on choices made in low-income countries, as we shall learn in the next chapter.

Further Reading

Trevelyan, J. P. (2014). *The Making of an Expert Engineer.* London: CRC Press/Balkema - Taylor & Francis, Chapter 13.

Chapter 18

Sustainability

Until this century, engineers designed for an infinite world: water was always available, and waste could be freely discharged. Now we have to work within finite planetary limits. Your job, as an engineer, is to help re-engineer our human civilisation as we transition to genuine sustainability. Engineers are some of the most influential people in making this happen.

Until recently, few if any companies would invest in engineering to improve environmental sustainability without a quantified business case: profitability had to be guaranteed. Now, most leading companies are demanding that their engineers find sustainable solutions, reducing or even eliminating net greenhouse gas emissions, improving energy efficiency, and minimising natural resource consumption, while maintaining preferably increasing profits. These companies have found that people will pay more for products produced in a genuinely sustainable enterprise, and they also know that governments are likely to lurch unpredictably to tighter environmental regulation with every natural disaster linked with climate change. Longer-term economic sustainability means anticipating these changes and taking advantage of profitable opportunities created by companies that don't manage to keep up.

Even though politicians are reacting slowly to this new reality, others realise that we have to change. Banks are increasingly anxious to provide finance for projects that respect sustainability goals. They know that if the world's media learn of toxic waste discharges from community activists armed with low-cost, yet sensitive instruments purchased online, it won't be long before reporters ask who provided the finance. They fear consumer boycotts in wealthy countries that can damage their reputation and limit their ability to raise finance for new projects. Banks have already, in effect, cut off finance for coal-fired electricity generators because they insist on loans being repaid in just a few years. They reckon it is unlikely that these plants will be allowed to operate after a few years without severe restrictions.

Unless you were very fortunate, your engineering school has not prepared you very well for these changes because it's difficult for faculty staff to keep up while an understanding of human influences and behaviour is critical for success. Most engineering schools provide little if any teaching on how behaviour and technology both shape sustainability. Human decisions influence sustainability everywhere, whether to turn an air conditioner on or off, buy a car, or decide on a building design.

Now it is up to you, as an engineer, to join the effort to achieve this huge transformation. Mostly it will be a long series of small improvements over the next three or four decades—your entire working career. It's the best time in generations to be an engineer; the world is increasingly desperate for your services. However, you need to be able to deliver results, something that you can't learn at university. That's why this book is so important.

One aspect of this transformation will be eliminating net greenhouse emissions: CO_2 (carbon dioxide), CH_4 (methane), and other gases.

Climate change

The essence of the December 2015 Paris climate agreement was to sufficiently limit greenhouse emissions to cap global warming at an estimated 1.5°C, if possible, and at no more than 2°C. Delegates also asked the Intergovernmental Panel on Climate Change (IPCC) to prepare a special report on the impacts of global warming of 1.5°C above pre-industrial levels and a comparison with 2°C of warming (Figure 18.1).

The findings revealed that 2°C of warming would bring unacceptable risks of irreversible damage to global ecosystems. They recommended an earlier cap on emissions to restrict warming to 1.5°C.

The clock (Figure 18.2) shows where we are today; we have to slow the clock before it reaches midnight to give ourselves more time to adjust.

This requires a rapid reduction in CO_2 and other greenhouses gas emissions into the atmosphere, as shown in Figure 18.3.

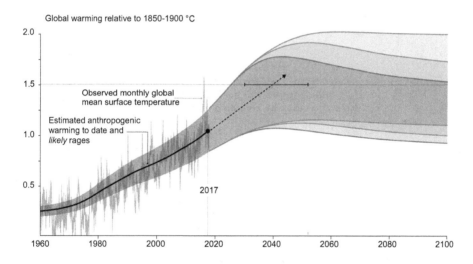

Figure 18.1 Expected climate warming, showing a range of warming from different climate models as different preventative measures are gradually put into effect.[1]

1 Figure SPM.1 from IPCC, 2018: Summary for Policymakers. In: Global Warming of 1.5°C. An IPCC Special Report on the impacts of global warming of 1.5°C above pre-industrial levels and related global greenhouse gas emission pathways, in the context of strengthening the global response to the threat of climate change, sustainable development, and efforts to eradicate poverty [Masson-Delmotte, V., P. Zhai, H.-O. Pörtner, D. Roberts, J. Skea, P.R. Shukla, A. Pirani, W. Moufouma-Okia, C. Péan, R. Pidcock, S. Connors, J.B.R. Matthews, Y. Chen, X. Zhou, M.I. Gomis, E. Lonnoy, T. Maycock, M. Tignor, and T. Waterfield (eds.)].

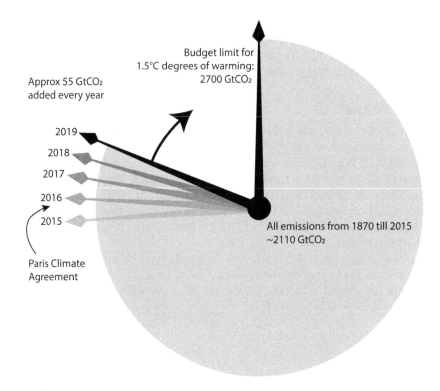

Figure 18.2 Count down to 1.5°C warming. When the minute hand of this clock reaches midnight, we will have discharged enough greenhouse gases. We can slow the clock by reducing emissions.

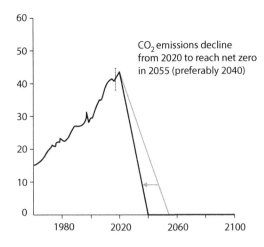

Figure 18.3 IPCC recommendations for emission reduction, reaching zero net emissions by 2045. Otherwise, greenhouse gas capture from the atmosphere will be needed as part of the solution. As yet, we don't have large-scale solutions for that.

Energy efficiency—using less material and energy to achieve desired outcomes—can provide as much as half of the necessary reductions, saving everyone money and effort. The remaining reductions will come from new methods that eliminate greenhouse emissions completely.

It should be noted that global warming targets are based on a 50% probability, and there is still a wide range of uncertainty in climate models. Normally, engineering decision-makers would regard a 50% chance of staying within a given environmental limit as an unacceptable risk; they would instead require at least a 99.99% chance. However, according to the best modelling available, we have already exceeded our atmospheric carbon budget for a 90% chance that climate warming does not exceed 1.5°C.

UN sustainable development goals

However, there are many other aspects of sustainability. In 2015, the United Nations adopted 17 'sustainable development goals' (SDGs) that cover all aspects of human civilisation and our natural environment, such as ensuring that plastic waste is biodegradable, cleaning rather than polluting air and water, reducing consumption of non-renewable resources while improving life for everyone. These are projects that will increasingly attract investment finance because they help us achieve SDGs. In wealthy countries, the emphasis will be on reducing resource consumption, recycling waste into useful products, and making more use of biological materials from renewable sources that can be reabsorbed into the natural environment. In low-income countries, the focus will be on improving productivity as well, which takes us back to the definition of engineering at the start of this book: enabling people to do much more with less. Many of these solutions will eventually be cheaper and safer; we will wonder why we didn't change sooner (Figure 18.4).

The challenge for many engineers will be to help clients take a leap of faith into this new world, rather than simply choosing the lowest cost option in accordance with the minimum permitted environmental standards.

Figure 18.4 The UN sustainable development goals.

Overcoming resistance to change

What if the company does not adopt this sustainability imperative?

Engineers face similar challenges with health and safety standards, particularly in some Middle Eastern and Asian countries. Owners are keen to minimise what they see as an unnecessary expense, especially when their competitors adopt similar practices.

These situations can present difficult ethical choices for engineers. When they are aware of practices that, even if legal, present serious health or environmental risks, then there is an ethical responsibility to inform the local community to help it realise what is happening. However, disseminating information that their employer regards as commercial secrets could lead to dismissal or even prosecution. In some regions, retribution can occur through non-legal means. Journalists can simply be beaten up by thugs, and families may be threatened when powerful people want to prevent further disclosures.

Apart from resigning, there are other options for engineers in these situations. Sometimes there are solutions where technical or commercial practices can be changed, providing large payoffs for the firm, the community, and the planet.

Here's an example. In the early 2000s, a Brisbane engineer at a food and beverage processing plant saw an opportunity to recycle wastewater...

> It took ages, but we finally convinced our management to install a reverse osmosis plant so that we could recycle process water, enormously reducing our consumption and also significantly reducing biological nutrient discharge into the waste stream. Both of these were significant gains for environmental sustainability. It took several years, because the management would not compromise on seeking a 25% return on investment. We tried several ways to make the necessary business case before it was finally accepted. Sometime after we installed the reverse osmosis plant, the city water supply was drastically curtailed because of drought. Our competitors had to close down their production lines because they were consuming too much water; they had to import products instead and sell them at a loss to maintain their market presence. We were able to continue at full production and made huge profits, repaying the reverse osmosis investment by several hundred percent. In retrospect, we should have based our case on the risk of water restrictions—it would have been accepted much more readily.

This account and many others like it show how unpredictable but foreseeable events can completely change commercial priorities, and it demonstrates how an engineer can argue for lifting health, safety, and environmental standards on the grounds of risk management, or even opportunities for large profits. Pursuing these options requires sustained persistence because it takes time for most business owners to recognise the commercial opportunities. Figure 4.1 is useful for understanding this reluctance. It takes time for someone who has not seen the dog in the right-hand picture to notice it. Many people will never see it without help.

Most business decisions assume that the economic conditions today are going to persist in the future. However, recent history demonstrates that this is a false assumption. Inflation in Australia today is around 1.7%. In the 5 years after I graduated, inflation increased from 3% to 18%! Housing loan interest rates today are about

4%, yet in 1989 they were 17.5%. Today's oil price is about USD 20, down from USD 60 when I started writing this book. Yet it rose from USD 25 in 2002 to USD 145 in 2008 (Figures 18.5–18.7).

Most engineering projects take around 5 years to complete from initial discussions, and these graphs show how much has changed in recent times. In addition, we can expect more regulatory changes in the coming decades. Disasters such as the recent bushfires in Australia and the Coronavirus outbreak will force governments to change policies. Therefore, it makes sense to anticipate change; projects designed for sustainability and resilience in the face of economic, regulatory, and environmental changes will provide more consistent investment returns for their owners.

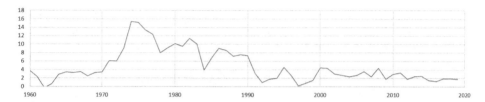

Figure 18.5 Australian retail price inflation. (World Bank data, accessed April 10, 2020.)

Figure 18.6 Interest rates in Australia (Data from Reserve Bank of Australia). Note that policy changes have altered the relationship between the overnight cash rate and the housing interest rate over time (Cash rate data from https://www.rba.gov.au/statistics/historical-data.htm, interest rate data from https://www.loansense.com.au/historical-rates.html, accessed April 10, 2020).

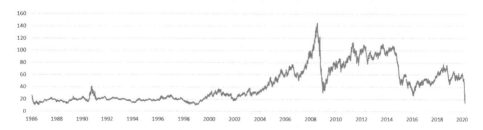

Figure 18.7 US crude oil price since 1987. A few days after creating this graph, the price was briefly negative. (http://www.eia.gov/, accessed April 10, 2020.)

Renewable energy

One of the greatest changes influencing engineering projects will be energy supplies. For more than a century, engineers have relied on fossil fuels and electricity grids to provide cheap energy on demand at stable, predictable prices. Now, the cheapest energy sources in Australia and many other countries are solar and wind-generated electricity. However, supplies are variable: the price can vary by a factor of 1,000 or more in just a few hours. Therefore, smart engineers are designing systems to take advantage of cheap solar electricity in the daytime, when prices can be zero, or sometimes even negative. One option is to design processes to operate at variable rates, slowing when energy supplies are reduced and prices rise. Another is to design processes to store cheap energy when it's available, sometimes in a different form. For example, a desalination plant can operate at full speed when cheap solar electricity is available, and store excess water in reservoirs for times when electricity prices are high. The stored water is equivalent to stored electric energy. Electrolysers can produce hydrogen and oxygen from water. The hydrogen can be transported long distances as liquid ammonia, and special membranes allow the hydrogen to be released as a fuel when needed.

Efficiency gains, new ideas, or behaviour change?

Efficiency gains will provide easy improvements with economic gains from energy and material savings. However, these alone will not be enough.

New ideas, often coupled with old ideas that have been around for centuries, can yield much larger improvements. Solutions like electric throw rugs for winter warmth, or fans and personal air conditioners for summer cooling, can provide comfort in extreme temperatures, using far less energy and material resources than heating or cooling entire buildings. Our current inefficient building heating and cooling systems consume as much as 30% of the world's energy, yet they provide comfort for only a minority of people on the planet. Once we leave behind the need to heat or cool a whole building, older low-cost construction methods become attractive once again, with more modest material requirements.

Ultimately, however, sustainability also relies on human behaviour. Common, shared resources such as the atmosphere, oceans, underground water reservoirs, and even forests and pasture must be managed cooperatively. Government regulation helps influence behaviour, especially if monitoring is feasible, enabling enforcement and deterring selfish behaviour. Engineers have a crucial role, deploying distributed networks of sensors coupled with satellites and global communication systems to monitor shared environmental resources. Even in the absence of effective governance, a reality in most low-income countries and sparsely populated regions, accurate monitoring of the state of the natural environment can provide data needed to motivate community-level responses and enable global networks of non-government organisations to target rogue players with effective sanctions. For example, the web site earth.nullschool.net and many others provide increasingly useful data on pollutants such as SO_2. Emissions of CFC refrigerants can also be monitored from space.

Mobile phone systems, especially in low-income countries, have become a trusted channel for financial transactions. As long as we can predict the capacity of shared resources, therefore, mobile networks with sensor networks could provide the means to govern access to shared resources. Mobile networks have proved to be remarkably effective without the need for expensive social institutions, such as extensive policing, to enforce payment by delinquent users.

We now know that Australian aboriginal people managed a sparsely populated continent with relatively infertile soils for at least 50,000 years using controlled burning. Their remarkable sustainability perhaps reflects their culture. Their notion of 'country', the area of land occupied by a tribal group, is very different from other cultures. 'Country' includes not only the land, sub-surface and air above but also the human inhabitants and animals. Any disturbance to 'country', therefore, is also a disturbance to people: it is a single organism or system. In most cultures, particularly our West European culture, the 'environment' is something that is distinct from humans. We talk of polluting the environment and miss the connection that we are simultaneously polluting ourselves.

Of course, it would be quite unrealistic to wait while we change human culture to embrace such connections: we have to act in the next two decades, so engineering solutions will be critical.

Opportunities

While new technologies can help, and in some cases will be essential, it's important to recognise that new technologies typically take 30–40 years from initial demonstrations to widespread adoption. Therefore, our sustainability transformation will mostly rely on existing technologies. There is a huge variety to choose from: all I can do in this final section is mention a few interesting examples.

In pursuing sustainable solutions, we cannot ignore today's inequalities: commercial solutions that reduce or eliminate inequalities (mobile phones, for example) are likely to be more sustainable than solutions that require wealth transfers through government agencies to offset any resulting wealth inequalities.

Access to piped water today relies on manually read meters and centralised government agencies to enforce payment. In low-income countries, piped water networks are mostly in a chronic downward spiral where contamination and poor service quality cause reluctance by users to pay their bills. Low revenue collections straining finances and low engineering productivity combine to undermine maintenance, resulting in more contamination and lower service quality. Users are forced to adopt expensive alternatives to obtain safe drinking water, such as purified water in 20 L bottles. The result: safe drinking water across South Asia can cost 10–30 times as much as in wealthy countries like Australia.

Alternative systems incorporating sensor networks and mobile phone payment systems should be commercially feasible and provide safe drinking water at a similar cost as in wealthy countries, potentially a huge improvement and commercial opportunity.

Non-sewered toilets can avoid the current practice where huge quantities of purified water are used just to transport human waste to treatment plants. Several new technologies are being tested and, optimistically, could be deployed in the next decade on a large scale.

While refrigeration and plastic packaging has enormously reduced food wastage in wealthy countries, we now have a plastic pollution headache. Low-income countries have to grow far more food than they eat because of high wastage in storage, processing, and distribution. The obvious solution is biologically friendly food packaging technologies that provide appropriate food preservation without long-lived pollution.

Pay-as-you-go technologies relying on mobile payment systems could enable farmers and small businesses to acquire expensive refrigeration systems that can further reduce food wastage. Bank finance taken for granted in wealthy countries relies on highly skilled people working in costly networks of retail bank branches, and there are not enough highly educated people to provide these services in low-income countries.

The most attractive opportunities to make technological improvements often become apparent only when you can think about the wider context in which technologies are used. Understanding cultural, economic, social, and governance factors, and trust in technologies, can lead to transformative innovations. However, this ability to think 'outside the box' must be complemented with the consistent application of systematic engineering methods, to ensure that all the technical details are resolved so that the expected benefits are realised.

You now have most of the knowledge you need to grasp these amazing opportunities. The one resource that will always be in short supply is your personal time; managing that is the subject of the next chapter.

References and Further Reading

Hardisty, P. E. (2010). *Environmental and Economic Sustainability.* Boca Raton, FL: CRC - Taylor & Francis.

Trevelyan, J. P. (2014). *The Making of an Expert Engineer.* London: CRC Press/Balkema - Taylor & Francis, Chapter 12.

Chapter 19

Time management

One of the most influential factors in an engineer's success is time management. As an engineer, you will have many competing demands on your time. Like many engineers, you may feel that most of these are 'interruptions' or other demands that impede your 'real engineering' work.

Here's how Leslie Perlow, author of the book *Time Famine*, described this:

> I found that engineers distinguish between 'real engineering' and 'everything else' that they did. They defined real engineering as analytical thinking, mathematical modelling, and conceptualising solutions. Real engineering was work that required using scientific principles and independent creativity. It was the technical component of engineers' deliverables that utilise the skills the engineers acquired in school. As one engineer summed it up, 'real engineering' is what I thought I was hired to do. In contrast, 'everything else' translated mostly into interactive activities.

Engineers describe these interactive activities as disruptions to their real engineering although interactive activities are equally critical for the completion of an engineer's tasks. Most of the time, these social interactions are unplanned and spontaneous, so they can seem like interruptions.

This can lead engineers into a vicious work-time cycle. In this cycle, time pressure (to get a product to market/to get the project completed) leads to a crisis mentality that results in individual heroic behaviour causing constant interruptions to others, adding to the time pressure and crisis mentality. 'Fire fighting' is the term used by many engineers to describe this, and it's ever-present for many, if not most engineers.

You can learn to break out of this vicious cycle.

Evaluate your own time management by taking the questionnaires at this web address:

https://jamesptrevelyan.com/2015/08/21/time-management-for-engineers/
   (or https://wp.me/p51WXF-3w).

Unless you scored at least 50 or more on each questionnaire, you can probably improve.

Understand daily physiological patterns

We all have different sleep patterns. Some people sleep for only 3 or 4 hours a night, while others need 8 hours. People who sleep only a few hours in the night often have the ability to take short 'cat naps' during the day: short sleeps between 10 and 30 minutes at a time.

Together with our sleeping cycle, we also have better and worse times for intense cognitive (thinking) work during the day. Some of us, including myself, can concentrate better on demanding work early in the day. If I have a challenging piece of work to do that demands concentration, the best time for me is between 7:30 AM and 10:30 AM. I also have another time during the day when I can concentrate better—between about 4:30 PM and 6:30 PM. Other people, often called 'night owls', work better in the evening and even into the early hours of the morning.

Try to figure out your best times for work that requires sustained concentration. Use other times for inter-dependent tasks that require interactions with other people and make people feel welcome when they seek your attention.

It's important to understand that our brains need biochemical energy supplies, especially for learning. Your biochemical energy supplies vary throughout the day. For some time after intense exercise, they will be depleted. Digesting certain foods, particularly meat in combination with carbohydrates, also demands biochemical energy, reaching a peak around 2 hours after eating. That's when you tend to feel sleepy after a heavy meal. Food consisting mainly of whole grains, such as oats or whole-grain bread, takes longer to digest and places less of a demand on biochemical energy supplies. However, we're all different in our responses to food and exercise.

Try to develop your awareness about your physiological state at different times during the day, and figure out when you perform at your best. Try to reserve these times for tasks requiring mental concentration.

Classify tasks

We can characterise some of the different tasks faced by engineers.

Extended thinking/intensive tasks demand concentration. Interruptions can easily disrupt this kind of work because we often rely extensively on short-term working memory: memories that persist for 10–20 minutes. Even a brief interruption can disrupt our short-term working memory, and it can take 10–15 minutes to resume the interrupted task. Studies have shown that in the presence of frequent interruptions up to 70% of computer tasks remain uncompleted at the end of a working day—we simply forget to return to them.

There are many other tasks that rely much less on short-term memory or take only a short time to complete. An example is filling in a form or writing simple replies to incoming messages and queries.

Social interactions are essential for technical collaboration but often present themselves as unwanted interruptions if they coincide with extended thinking tasks. Research has shown that technical collaboration relies on socio-technical interactions that take up 60%–80% of working time for engineers.

When classifying tasks, also learn to think about how each task contributes to commercial and social value generation, as explained in Chapter 15. While it is often difficult to quantify value generation until tasks have been completed, expert engineers often ask themselves, "How is this going to generate enough value to cover the cost of time and effort?" Helping other people create value through their own work is often the most successful way to do that.

Adapt your schedule

Learn to adapt your schedule according to your physiological state and the demands of the work that you have to accomplish. Always remember to pre-allocate some spare time for the tasks that you don't anticipate. Observe your ability to predict how long it takes to get things done, and then learn to compensate with your schedule. If you find it takes you twice as long to get things done as you expected, allocate twice as much time as you think you will need.

If you don't have enough time to allocate, plan ahead. Let people know what you can deliver and when. If you can't deliver, it's much better to say so ahead of time so that everyone can adapt. Some simple advice: always try to deliver more than what other people expect, earlier than expected, and allocate 50% more time than you think you need.

Use an electronic diary or calendar to plan your time. Remember, the more you have in your calendar, the easier it is to tell others that you're too busy to help out immediately unless you really want to assist.

It is wise to disable all email notifications and decide your own schedule for checking your inbox. Create the expectation among colleagues that they should either visit you, telephone, or send a brief text message if they need an immediate reply to an email.

Keep records

There is a universal principle in engineering: you can't manage something that you don't measure. The essence of time management is an awareness of how you spend your time; comparing the actual time you spend with your original predictions. That's the only way to improve, by observing your actual prediction performance and modifying your predictions in the future.

Why is this so important for engineers?

Obviously, it helps to be able to deliver work when other people expect it. However, what's more important is that, as an engineer, you will be relying on lots of other people to perform skilled work to help you complete your tasks. This is inter-dependent work requiring technical coordination and project management, and it typically takes at least half of the engineer's time, just organising the work to be done and monitoring what other people are doing.

One of the most important skills, therefore, is developing the ability to accurately predict how long it will take other people to get technical work done. You will depend as much on that as managing your own personal time.

From time to time, keep detailed records and observations of what happens during each working day. Identify times when you are more likely to be interrupted and other times when you can safely focus on time-intensive thinking tasks.

Schedule major tasks

Many people only schedule meetings in their calendars. This is a missed opportunity for time management.

The largest component of working time for most engineers is technical collaboration. It's tempting to see this as 'non-technical' engineering work, but research shows that:

i   The concepts engineers discuss with others are technical and rely on technical understanding, though much of the time they are unconscious understandings involving tacit knowledge. See Chapter 10.
ii Technical collaboration is essential in engineering: without it, engineers would never achieve much.

Learn to see this as core engineering work and allocate times during the day when the people you collaborate with are present or most likely to call.

Allocate other times in your calendar for extended thinking tasks—when you are least likely to be interacting with other people—so as to minimise the chances of interruptions. Make sure that other people sharing your calendar can see that you will be busy at these times. Agree on a signal with your colleagues that indicates you want to avoid unnecessary interruptions—for example, wearing earphones (even if you don't have music playing). Switch off your phone or put it on silent. Close chat windows and all other distracting applications on your computer screen. Close your email application.

Allocate time to help others

As a rough guide, allocate 30–45 minutes daily for every person you're expected to help or supervise on a routine basis. By scheduling work on your major tasks around these other commitments, you will not see requests for help as interruptions. Over time, other people will sense a positive and welcoming attitude, which will help build respect.

Say "no" by saying "yes"

It is a good idea to schedule time for extended thinking tasks many weeks ahead, even if you're unsure what you will be doing.

When someone asks you to collaborate on some task (i.e., coordinating your contributions), it's handy to have a diary that seems packed with commitments.

Younger engineers, in particular, find it difficult to say "no" to these requests because almost everyone prefers to be seen as helpful by others, particularly by senior people.

With a packed diary, one can safely reply, "Yes … I would love to do that. How about … (choose a time several weeks ahead)" while gazing intently at one's calendar, trying to find a place to slot in the required task.

The first part of the sentence conveys the impression that you are keen to contribute. The second part generates respect that you are so fully booked ahead that even saying yes seems like granting an immense favour.

Defer or delegate: documentation and filing is the key

The final strategy to use on tasks that you cannot complete is to delegate them to other people—or simply ignore them.

Delegation is only possible on short notice if the task, and what you have completed so far, is documented in a way that someone else can easily see what needs to be done next, and what is needed to complete the task. They will also need to know where to find the necessary information on shared cloud storage and the contact details of everyone else involved. Routinely documenting your work in this way is one way to manage your time proactively.

Make sure you set aside time each day for email correspondence: 1–1.5 hours should be enough.

Ignore all but the most critical emails until this scheduled time.

Set your email default options such that when you click 'Send', your email is placed in your 'outbox' until you click 'Send all emails now' (i.e., disable the 'immediate send' option). If your email system does not allow this, make a habit of placing all outgoing messages in your 'drafts' folder. Then, once or twice a day, review and send all the waiting draft messages.

Following this process will help you quickly decide how to respond to incoming messages:

a   Open an email.

b   If it requires a brief reply, such as "Thank you, I have noted that," write the response and dispose of the email by filing it in a designated folder. Deleting it is usually not an option: every aspect of engineering requires an audit trail so you can establish that you considered the issue with the level of 'due diligence' that a normal engineer would consider appropriate in the circumstances.

c   Consider the consequences of not replying at all.

i   Has the email also been sent to other people who can be counted on to reply?

ii   Does it request a specific response from you?

iii   Does it contain information that other people cannot explain to you later?

If the answers are Yes | No | No then you can almost certainly ignore the email. If you still cannot ignore it, make a time in your calendar to read and reply to the email, and send a brief reply to let the sender know when they can expect a reply.

Remember that, almost certainly, your reply will trigger another email from the sender.

Decide when the sender really needs a response and schedule your reply one or two days before then. Once people get used to this practice of yours, and they know that you will send a carefully considered reply, they will be careful about asking you and will try to provide all the information you need. If you are known to reply quickly to emails, then you will almost certainly receive more emails, often because other people have not read your replies carefully.

i   Is it best answered by someone else?

If yes, then approach this person, preferably face to face or by phone, and ask them to compose the reply. Reserve a 15-minute time slot in your diary to follow up after a few days to check that the issue has been appropriately handled.

Consider using rules in your email software to file routine emails automatically in a folder, such as newsletters and standard email circulars. That way they don't clutter your inbox, yet you have the emails filed away should you need them.

Unforeseen disruptions, avoiding overwork

Regardless of your time management discipline, there will be major disruptions caused by unforeseen events, both in your professional and family life. For example, a colleague may suddenly leave, have a major illness, or be transferred to another task, and you may be asked to take on his or her responsibilities.

As explained above, maintaining up-to-date routine documentation of all your on-going work activity means that you can quickly delegate tasks to others, if needed, while you handle the emergency.

Some managers expect instant updates with reports on issues that you're handling. That's another reason to keep the documentation up to date in anticipation of such requests. You will always have something ready to send. Usually, by the time the recipient has worked out that the report is not exactly what they needed, you will have updated your documentation and will have those issues covered—for the next urgent request.

If you plan and utilise 100% of your time, you will never catch up after such a disruption.

Therefore, your planned work should require substantially less of your time, allowing you to cope with disruptions. Many engineers are tempted to fill all their time or even work long overtime hours, believing that this will advance their career prospects. However, research and experience show that fatigue induced by working long hours has two pernicious effects. First, you will make more mistakes, and the time needed for you (and others) to recover from those mistakes will outweigh the advantage gained by working extra hours. Second, your productivity will decrease and your ability to assess whether the work is adding value will decrease, resulting in you having to complete more work than needed, often called 'overwork'.

By questioning the value added by each of your tasks and responsibilities, you can aim to maximise value per hour worked.

By planning a schedule with a proportion of time free, you will be able to take advantage of unexpected opportunities without compromising ongoing work.

Following a systematic approach to time management will help you overcome the 'time famine'. However, there are several other frustrations that can spoil your day; the next chapter explains how to overcome them.

Chapter 20

Frustrations

Engineers experience many frustrations.

Completing any engineering degree course is tough; at most universities, students consider it one of the most difficult areas of study to pursue. As many engineers soon find out, an engineering degree course provides a great foundation, but it may not lead to a great career. Many engineers find themselves in dead-end jobs with little chance of being promoted; they may even see non-engineers being promoted faster. Often, they see few ways to apply their technical skills and begin to question what all the hard work they put in to become an engineer was actually for.

Why are so many engineers so ill-prepared for engineering workplaces? Part of the reason is that most of us, long before graduation, developed expectations about what engineers do and how other people behave based on popular myths and stereotypes. Many students graduate thinking that engineering is a hands-on, practical career. Others imagine they will be spending most of their time solving complex technical problems or creating innovative designs. They imagine that 'real engineering' is all about design, calculations, and thinking about technical issues.

This is why many engineers feel frustrated when their work lacks the kind of technical challenges they have imagined that 'real engineers' spend their time on. Working with other people can also bring a different set of frustrations.

Learning how to avoid or work around these frustrations will help make your work experiences far more enjoyable and productive (Figure 20.1).

Frustration I: Working hard is not getting me anywhere

Many engineers, when faced with a challenge, work longer hours in the belief that working harder will improve results. Often it only makes matters worse because it can reduce your opportunities for social interactions on which collaboration depends. Also, as explained at the end of the last chapter, fatigue from long hours will reduce your productivity and possibly result in mistakes that take you and others far longer to fix.

When collaborating with other people, success depends on everyone succeeding. Helping others succeed, therefore, is necessary for success in engineering.

Interdependent, collaborative working requires continuing engagement. Technical ideas that emerge through engineering work almost always will be implemented by other people. It is essential that the ideas are preserved sufficiently intact so that

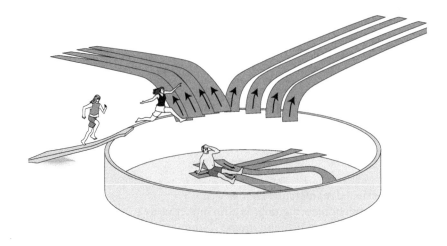

Figure 20.1 Off to a flying start? The steep inclined ramp on the left side represents your undergraduate engineering education. Getting an engineering degree is just one obstacle that lies between you and the wonderful career that lies ahead. It is a steep climb, but it gets you over the wall and into the ring. However, many engineers experience a hard landing, like the fellow in the picture. Some careers can even seem like dead-ends with no way out.

predicted technical and commercial performance is achieved by the solutions created through all this work. In addition, engineers themselves seldom know enough to completely specify the technical intentions. Therefore, others steadily elaborate on the original ideas, adding their knowledge and insights. Engineers, however, must monitor implementation to ensure that the original intentions are not compromised through these elaborations. It is not enough to hand over a set of instructions and leave it to others to implement them.

Of course, organizing and collaborating with other people takes time and effort. There's a joke that explains this well: "The IBM definition of a man-year is 365 programmers all trying to finish the one project in a single day." Often engineers exclaim in frustration when collaborating with others: "It would have been quicker to do it myself!" Maybe, except that no engineer can do everything.

Frustration 2: I can't get a job without experience and advertised jobs require experience

Many engineering graduates face this frustration, especially after years of study and possibly hundreds of job applications. It results from repeated advice for students in schools and universities on how to apply for advertised jobs. Almost all jobs in schools and universities have to be advertised because of government regulations. Therefore, teachers simply may not know that most jobs are never advertised.

Why do nearly all job applications require experience?

The reason is simple: nearly all job applications are written with the hope of finding a replacement engineer with the same level of experience as the one that just left. This habitual practice overlooks the reality that the engineer who just left probably accumulated experience while performing the work and may have started with little relevant experience. Employers hope they can recruit an equally experienced engineer. If not, they will probably recruit a novice engineer with minimal experience and, hopefully, provide enough support for the novice to rapidly develop relevant experience on the job.

Therefore, the way around this apparent obstacle lies in Chapter 3.

Frustration 3: Admin, meetings, accounts, and procedures: this is not what I was educated for

"Why do I have to spend so much time on administration, emails, filling in forms, going to meetings, and all that non-technical stuff?"

This is one of the most common frustrations I have seen among engineers. I experienced it myself many times throughout my career.

These aspects of engineering are essential because they are the means by which people collaborate in a technical enterprise and, as we have already seen, collaboration is an essential part of all engineering work.

Engineering education focuses almost entirely on the small white parts of Figure 20.2 with dashed outlines: analytical tools to support technical performance prediction, along with some basic design skills. Delivering results for clients is hardly mentioned. Yet this is what clients pay for, and engineers spend most of their time doing. Even in the upper thread, much of the work involves extensive collaboration with other people, many of whom are not engineers. Yet, engineering schools seldom mention collaboration, the essential human social interactions at the core of engineering. As explained in Chapters 9–12, most access to technical knowledge comes through social networks.

No wonder so many novice engineers complain that "This is not what I was educated for!"

Therefore, as one might expect, many engineers find much of their work difficult, time-consuming, and frustrating, simply because they have not learned effective methods of collaboration or how to deliver practical results for clients.

Figure 20.2 shows how many engineers see large parts of their work as 'non-technical' because it does not align well with their engineering studies at university. Yet, when asked, engineers readily admit that engineering knowledge is essential for these 'non-technical' aspects. That is why they are better described as 'socio-technical' work: collaboration requires social interactions in which specialised technical knowledge contributes to success.

This book will help you learn to effectively collaborate using socio-technical performances like technical coordination. When you develop appropriate skills, you will find that collaboration can be easy and enjoyable and no longer frustrating. So many aspects of delivering practical results for clients depend on the particular industry and discipline you work in; there is too much to present in a single book. However, the technical collaboration skills you can acquire from this book will help you build your expertise to deliver results for clients.

Now take another look at Figure 2.2, which shows engineering practice supported by three legs: engineering and business science, perception skills, and tacit ingenuity.

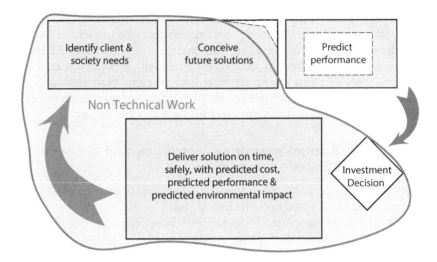

Figure 20.2 Non-technical work?

Most engineers graduate with only one leg at best. They have little, if any, technical foresight and planning ability and hardly any awareness of technical collaboration. No wonder engineering practice can be difficult and frustrating for so many graduates.

It is sad that so many engineers give up in frustration and leave the field for other occupations, often with extensive further education or workplace training courses. In many low-income countries, engineers become semi-skilled computer programmers, a job for which they also have little education, but one that often provides higher pay than they can earn as engineers.

Frustration 4: This job does not have enough intellectual challenges for me

Some other engineers expressed such frustration in this way: "I feel I am not exercising the technical skills I was trained for. I envy engineers who have jobs where they get deep into technical issues, like R&D."

This frustration is common among nearly all engineers. As students, our days were spent solving technical practice problems, so we imagined that we would be spending much of our time on technical work as engineers.

As engineers, we see the challenge of solving difficult technical problems as one of the most satisfying aspects of our work.

Engineering is much more about routine processes than solving difficult technical problems. Much of the routine has evolved as a way of avoiding the need to solve problems that have already been effectively solved many times before.

The best engineers I interviewed for this book all told me how much they enjoyed the intellectual challenges that engineering presents. They all explained, in different ways, how they had gradually learned that the most rewarding challenges come from finding effective ways to overcome human limitations and helping other people succeed despite their handicaps.

The engineers who expressed frustration at the lack of technically challenging work were still struggling to make that shift in perspective. Given that engineering depends so much on collaboration, it follows that improving engineering performance depends as much on overcoming human limitations as it does on technical advances. Even technical advances depend on people coming up with appropriate new ideas, which is very much a human process. So, in both respects, the ultimate intellectual challenges for engineers are about overcoming human limitations.

Learning a little about human behaviour should help you to appreciate and enjoy the intellectual challenges in finding ways to overcome human limitations. Think about the complexities of commercial aviation today and the amazing safety record achieved over many decades. This has been achieved only by rigorous intellectual consideration of human behaviour in the context of complex technical systems.

Frustration 5: Has this been done before?

An engineer told me this in an interview:

> The only two times I really enjoyed myself in my career were when the clients forgot to ask me 'Has this been done before?'

Many engineers told me about the frustrations they have experienced because their clients did not allow them to try anything new. This is normal. Most clients see innovation as inherently risky, and they will often choose more expensive solutions just because they have worked before. Chapter 15 explains how eliminating risks makes a project more valuable. Chapter 11 in *The Making of an Expert Engineer* explores this idea in a greater level of detail.

However, it is possible to work around this frustration. We can learn much from C. Y. O'Connor, an Irish engineer who planned and oversaw a remarkable engineering project in Western Australia in the 1890s. He proposed a pipeline to transport water from Perth on the coast to Kalgoorlie, 550 km inland, in a dry desert of salt lakes (Figure 20.3).

At that time, the world's greatest water pipeline had been constructed to carry water from a reservoir in England's Lake District to Manchester. The pipe was about 250 mm in diameter and about 100 km long. O'Connor's pipeline was to be 785 mm in diameter and about 530 km long, with a total hydraulic lift of 480 m, using a completely new method for manufacturing pipes and joining them.

Engineers today have a term for this: step-out. It means the degree to which an engineering undertaking uses untried technology or goes beyond the limits of what has been done before. In today's engineering terms, O'Connor's pipeline was not just a step-out; it was an unimaginably huge step-out in a place with virtually no established engineering capacity—a sparsely populated backwater at a distant extremity of the British Empire at that time.

In order to get the project accepted, O'Connor cleverly argued that he was simply constructing a series of 14 separate pipelines, each running from a pumping station to a holding reservoir, which was well within routine engineering capabilities at that time in terms of diameter, length, and hydraulic lift. What was new was that each of these pipelines would be placed end-to-end in order to achieve the required objective of bringing water from Perth to Kalgoorlie.

Figure 20.3 Charles Yelverton O'Connor. (W. A. Water Corporation Library.)

He correctly anticipated that the engineers hired by banks financing the project to assess its technical feasibility would focus on the overall scope of the project, rather than small technical details. Some of the major technical uncertainties lay in those details, however, and O'Connor ran a series of tests to ensure that his novel methods for joining the pipes would work.

O'Connor's foresight was remarkable. The pipeline was completed only 8% over the original budget, even though there were no maps for O'Connor to plan the details. It is still in use today, more than a century later. Read Chapter 4 in *The Making of an Expert Engineer* for more details on this project.

Frustration 6: Constrained by standards?

Many novice engineers (and engineering faculty as well) often think that codes and standards constrain designers: you have to comply with their requirements, thereby inhibiting creativity and innovation.

Expert engineers know that codes and standards represent accumulated engineering experience. Following codes and standards almost always saves time and reduces uncertainty, cost, and risk perceptions. Creativity lies in knowing how to work within the constraints imposed by codes and standards to achieve outstanding performance.

While codes (normally) prescribe mandatory practices and methods, standards usually provide guidance and recommendations, rather than strict rules. However, it is always sound practice to justify any instances where it is necessary to deviate from standards.

Standards, along with design guides and application notes in component manufacturers' catalogues, provide proven, simplified design methods.

For example, Australian Standard AS4055 is a simple method for calculating the maximum forces due to wind on a small building such as a house or shed. For a given terrain type (e.g., flat, undulating, hilly), with or without nearby equivalently sized trees to provide shelter, in a given area of Australia, the standard provides a measurement of maximum wind pressure, which might be given as 1.3 kPa, for example. The maximum force on a section of the roof can then be calculated by multiplying the maximum wind pressure by the area of the roof section. Different pressures are provided, both positive and negative, depending on the relative wind direction, for different parts of the structure of typical small buildings. The maximum wind pressures in areas of Australia affected by cyclones (hurricanes) are obviously much greater.

Everyday engineering practice in the commercial world often depends on people who are tired, bored, prone to forgetfulness, and anxious to get home.

Codes and standards enable engineers to routinely produce safe designs quickly and economically, with a much lower chance of making mistakes than if they were working from first principles.

Using standard methods directly reduces costs. However, there are much greater benefits because investors perceive less risk when engineers follow design codes and standards. That factor alone can greatly increase the value of an engineering project, as explained in Chapter 15.

Most organisations that issue codes and standards act independently of governments and must recover their costs from subscribers, despite the fact that they may carry an implied connection with governments. Standards can be expensive to purchase, even online.

Expert engineers can list nearly all of the standards relevant in their area of practice from memory and will be familiar with most of them.

Frustration 7: Yearning for hands-on work

Many engineers start out with the idea that engineering is a hands-on occupation, with plenty of opportunities for practical work with machines, circuits, computers, and construction.

In reality, most engineers rarely (if ever) perform hands-on work and are often required to ensure that any hands-on work is performed by technicians with appropriate qualifications.

Hobbies provide engineers with the best opportunities for hands-on work. If this is important for you, then start a hands-on hobby today.

Some engineers do get their hands dirty from time to time, of course, and these experiences can provide rich learning opportunities.

Frustration 8: I can't get other people to understand my ideas

During my research, many engineers reported frustrations associated with their apparent inability to influence others, let alone help them understand something that, for an engineer, seems glaringly obvious.

Overcoming this difficulty opens huge opportunities for engineers who, after all, can only succeed by influencing others to do their work differently or use different tools—in other words, teaching.

We often hear engineers characterised as 'nerds' with appalling social skills; one could easily dismiss this difficulty as a consequence of limited communication abilities.

Gradually, through my research, I came to realise that most engineers acquire several deep misunderstandings about communication and language. Many of those reading this may share some of these misapprehensions. I have seen how some engineers have learned to be remarkably effective and persuasive because, somehow, they overcame these misunderstandings. You could too.

At the same time, as creating solutions for people, engineers must also be teachers. The essence of teaching is helping others understand ideas better. Therefore, the way to overcome this frustration is to learn from Chapters 7 and 8 in *The Making of an Expert Engineer*. Fortunately, our knowledge of human learning has been greatly extended in the last few decades. Old ideas, like lecturing people, something you probably associate with teaching, have long been surpassed by evidence-based education methods that really work. Learn some of these and you may find engineering to be much more rewarding.

There is a simpler place to start, however. Go back and re-read the chapter on listening and learn to watch for when your audience has stopped listening. If that happens, there is no way they're going to understand what you are talking about. Learn to listen to your audience first, to acquire their language, and identify the special ways they use English. Then try explaining your ideas using their language instead.

Frustration 9: This company is run by accountants

Or lawyers. They don't understand even the simplest ideas in engineering.

Money rules in engineering—most of the time. For some engineers, commercial necessity is an unfortunate one: they would prefer to do "what's right, irrespective of financial constraints." However, decision-making in most successful enterprises combines engineering, legal, and financial accounting priorities. Many engineers find it hard to understand the non-engineering perspectives.

Even though the influential accountants and lawyers in your firm probably speak in English, they use English in very different ways than engineers, making this frustration, in essence, a language issue.

Now, a question for you. What do you think an accountant means when they mention accrual accounting? I've only met two engineers who could provide an accurate explanation, yet accrual accounting is a basic accounting concept, even more so than conservation principles are for engineers.

The lesson here lies in language. If you wanted to go on a date with an attractive French partner for a romantic dinner for two, and the partner could not speak any English, you would most likely take the time and trouble to learn basic French (like 'Je t'aime.' – 'I love you.'), and you would probably take a phrasebook or electronic translator with you. Dinner would not be much fun without being able to share some basic and simple conversation.

Therefore, it follows that engineers must understand basic concepts and language that accountants use in order to have anything close to a productive conversation. It is not very helpful or productive to simply blame them for a lack of understanding about engineering fundamentals. Take every opportunity you can to talk with accountants and lawyers to learn their languages.

Frustration 10: They always cut the maintenance budget first

Most engineering schools delegate discussions on finance to a business school. That explains why engineers often describe finance-related frustrations such as this one.

When expenses need to be cut back, it is easy to postpone maintenance because, most of the time, there will be no immediate effect on operations. Also, maintenance is often described as a 'cost centre', setting it up as an easy target for cost-cutting.

It has become fashionable, at least in maintenance engineering communities, to adopt the term 'engineering asset management' or, especially in defence circles, 'sustainment.' Part of the reason is that maintenance has historically been seen as a lower status engineering activity, organised by former technicians who have earned engineer status through years of experience rather than a degree qualification. With the name change, there has been much more emphasis on planned activities such as inspections and scheduled equipment shutdowns, rather than repairs after breakdowns. This is a more economical approach for maintenance. Organising repairs after a breakdown is almost always many times more costly than planned maintenance to prevent breakdowns.

Arguing for maintenance spending requires that you first understand the risk appetite of the enterprise owners. In some ways, maintenance is like an insurance policy. Some owners are happy to accept a higher risk of failure and unplanned shutdowns. It might be worth asking whether they pay for regular servicing of their car, or if they wait for it to break down and risk an expensive repair bill. A risk-averse person would rather replace their car with a cheaper model than do without insurance.

An enterprise owner with a lower risk appetite is more likely to be persuaded by an argument that emphasises value protection from effective maintenance (point 12 in Chapter 15). In a more risk-tolerant enterprise, such as a small mining company or a highly profitable manufacturing enterprise servicing short-term market fashions, a more effective argument might be based on the risk of an extended production interruption during a period of peak demand for the product, thereby allowing other suppliers to grab market share.

Frustration 11: They are only interested in the lowest price

As engineers, we understand that the solution with the lowest lifetime cost is quite possibly not the one with the lowest purchase price. However, implicit in our assessment is a forecast of future operating and maintenance costs, as well as a possible disposal cost. Forecasting anything in economics is fraught with difficulty, so our implied or actual forecasts will not necessarily be accepted by decision-makers.

Like arguing for maintenance, working around this frustration is likely to be centred on understanding the risk appetite of the enterprise owners. Some business models rely on a large profit in an inherently risky market, such as anticipating short-term demand for fashionable products or minerals in a volatile trading market. With a short-term payoff, the enterprise perceives today's expenses as much more valuable than next year's. In other words, there is a high discount rate applied to future expenses, either quantitatively or implied in decision-making. To understand more about financial decision-making, investor risk perceptions, and other influences, read Chapter 11 in *The Making of an Expert Engineer.*

It is equally possible, particularly in a government or developing country context, that all procurement is handled by a centralised purchasing department, and their procedures require them to accept the lowest complying bid for providing services or products.

In this situation, it is often possible to shape the purchase decision with a carefully written specification or statement of requirements that eliminates solutions with low purchase prices and higher operating and maintenance costs. Read Chapter 14 to learn more about specification writing.

Frustration 12: Net Present Value (NPV) shows the project is fine—why don't they approve it?

NPV calculation lies at the heart of most engineering economics or finance courses and some entrepreneurship programs. It is a way of taking into account the cost of financing a project over time, especially when positive cash flow is achieved some time after the expenses of design, planning, construction, and commissioning. Unfortunately, most courses and textbooks end there. A decision on whether to proceed with a project usually depends on many other factors, not the least of which are the apparent risks for the investors that the project might fail, run late, cost more, or provide income well below expectations. Large companies usually have many potential projects to choose from. To understand more about financial decision-making and investor risk perceptions and other influences, read Chapter 11 in *The Making of an Expert Engineer.*

Frustration 13: My skills and knowledge are only valued in rich countries

This is a common and often unspoken frustration among engineers in low-income countries. When they encounter enterprise owners who seem uninterested in technical innovations, they imagine that businesses in wealthy countries are much more receptive to these ideas. They have seen many news reports and videos about innovation, and they imagine that it is much easier to find jobs working with advanced technologies in 'the West'.

Of course, most of the innovation videos on the internet are actually there to try to attract investment funding. As engineers in wealthy countries know all too well, firms there can be just as reluctant to invest in innovation—see Frustration 5 above!

Low pay does not help. However, the research for this book has shown that engineers who generate value for their employers and clients, over time, tend to be well rewarded, and they can earn as much or more in a low-income country as engineers can in wealthy countries. Valuable engineering performances can be even more welcome in low-income countries than in wealthy countries. However, it is harder to deliver these results. The reason for writing this book is to help more engineers learn how to do that and overcome the special challenges faced in low-income countries. See Chapter 17 for more on this.

Frustration 14: I would much prefer a job where I could do something to help people

Here, all I do is make someone else richer.
I've had enough of working for fossil fuel companies.

Or, in defence...

I want to work on something creative, not machines that destroy people.

Many engineers take time out from their careers to pursue humanitarian or disaster relief projects. Many progressive engineering firms help their people do this, knowing that they will learn new skills, enlarge their networks, and possibly return more motivated.

However, not everyone has these opportunities.

If not, then see if the following line of argument helps.

Fossil fuels will be needed by many people for decades to come. Engineers can help make the most efficient possible use of fossil fuels, minimising emissions, and preserving non-renewable energy resources. They can help reduce waste and pollution, converting waste into valuable byproducts.

Lots of small productivity improvements add up over time to make a big difference in peoples' lives.

Even in defence, as I have outlined in Chapter 15, weapons with destructive potential can provide higher levels of protection and can also deter violence by others. Re-read Chapter 15 to understand how engineering helps stakeholders, especially the local community. There may be lots of opportunities to build value in the community as part of your work. Go out and look for them.

Frustration 15: My emails go unanswered

Email is a great way to transfer information from one computer to another. It is not a good way to gain sufficient interest for the recipient to reply.

If you're waiting for a reply, try going to visit the person at their workplace. That is the single most effective way to elicit a response. Think of a place where you might encounter the recipient in person.

If you cannot visit, then try calling by telephone.

If neither works, for whatever reason, wait...

After enough time has passed for the recipient of your email message to feel embarrassed that they have not replied, try sending the email again, adding "Are you OK?" at the end of the subject line. Explain that you sent the email several weeks before, and provide the date.

If that doesn't work, think of people who might be able to gain the attention of the recipient and try contacting them with a request to talk with your original recipient.

Epilogue – next steps

By reading this book, you will understand more about people, yourself, collaboration, and engineering practice. It is just a start, but enough to enable you to join the transformation of our world, enabling people to do much more with much less effort, time, material resources, energy, uncertainty, health risks, and environmental disturbance. You will still be learning more about people decades from now.

Extending your engineering practice knowledge

In such a short book, inevitably there are important aspects of engineering practice that have been either simplified or omitted entirely.

The Making of an Expert Engineer, published in 2014, can provide you with a much more detailed understanding of different collaboration methods: informal teaching and learning, technical coordination, project management, financial decision-making, and negotiation. It provides a deeper understanding of the human language and social interactions that form the foundations of these methods. It also provides a detailed explanation that can help you understand financial decision-making as it applies to engineering projects.

Professional recognition

After the first 4 years of engineering practice, it is worth seeking recognition for your accomplishments.

The most common recognition in many countries is a Chartered Engineer, or a Professional Engineering Licence. In many countries, this qualification is a legal requirement for independent, unsupervised practice.

In most countries, professional qualifications are administered by special organisations such as professional licencing boards. Applicants require several years of supervised practice and will normally be required to provide full details of their professional achievements. In the USA, Canada, and some other countries, passing a professional practice examination may also be necessary.

There are international agreements that extend professional recognition to many other countries, such as Asia-Pacific Economic Cooperation (APEC) Engineer and the Engineers Mobility Forum (EMF). The International Engineering Alliance has been successful in gradually building a consensus for standardising engineering education. Work is progressing on standardising professional recognition across national borders as well.

However, engineering careers are extraordinarily diverse, and many do not require a professional recognition or licence at this time.

Technical specialist, generalist, or management?

Engineers face many turning points: times when they may have to choose different career paths.

An early one may be the choice between working in the field or on the shop floor of a factory or pursuing a more analytical career working mainly in offices. Experience with both is very helpful in opening up more career choices.

Traditionally, engineering careers reach a stage at which, it is thought, engineers choose to become managers or technical specialists. In some countries and enterprises, managers are not even considered to be engineers.

Another decision is whether to specialise in a particular discipline or field of engineering or to move into other fields with a broader range of knowledge.

Our research on several hundred engineers did not reveal any consistent career patterns. We found engineers who started on construction sites, went on to manage large construction firms, and then became specialised technical consultants. We found others who moved through many specialist disciplines throughout their careers. Others moved into finance and business administration.

We found some evidence that engineers working as designers were a little less happy with their careers than others, perceiving much of their work to be a waste of time since very few of their designs were ever built.

Mostly, engineers followed opportunities that arose in the course of their work rather than following a long-term career plan. Many were only able to appreciate their particular strengths when placed into unfamiliar roles that brought out the best in them, often with the help of supervisors and mentors who recognised their potential before they did themselves.

Importantly, we found that almost all the engineers we encountered were extremely happy with their career choices and none regretted being an engineer.

The best time to be an engineer

Now, more than at any time in the past, the world needs engineers who can improve productivity and create sustainable solutions. The opportunities are there for you to grasp them.

The next step is your choice.

Online Appendices

Additional online supplements to this book can be found on the book page at www.routledge.com/9780367651794. These include a professional engineering capability framework, a listening skills observation worksheet, learning to see by sketching and a checklist of project lifecycle cost categories. To access these free supplements, please visit the book page and click on *Support Material*.

I Professional engineering capability framework

This document provides a comprehensive list of workplace performances that a supervisor or mentor could expect a novice engineer to demonstrate at a professional standard in the first three years. The list is grouped by the sixteen Engineers Australia professional engineering competencies. The document is formatted as a checklist by which a novice engineer can accumulate evidence of his or her achievements and performances as part of a portfolio to support an application for professional chartered engineer status or license.

The document complements the material in chapters 9 – 20 in the book, providing additional learning exercises to be completed in the workplace and additional reading.

Each novice engineer should retain a printed or marked up PDF version of this document as a record of his or her achievements and workplace learning progress.

2 Listening skills observation worksheet

This provides a checklist to help identify listening strengths and weaknesses.

3 Learning to see by sketching and a checklist

A guide for improving seeing skills by freehand sketching exercises.

4 Checklist of project lifecycle cost categories

A checklist based on Australian Standard AS4536 to help identify costs that need to be identified to assess the complete life cycle cost of an engineered product.

Index

Note: **Bold** page numbers refer to tables; *italic* page numbers refer to figures.

Printed in the United States
By Bookmasters